WHAT? WHERE? WHEN? WHY?

AUSTRALASIAN STUDIES IN HISTORY AND PHILOSOPHY OF SCIENCE

VOLUME 1

WESLEY C. SALMON

WHAT? WHERE? WHEN? WHY?

Essays on Induction, Space and Time, Explanation

inspired by the work of

WESLEY C. SALMON

and celebrating his first visit to Australia,
September — December 1978

Edited by
ROBERT McLAUGHLIN

*School of History, Philosophy and Politics,
Macquarie University, N.S.W., Australia*

Springer-Science+Business Media, B.V.

Library of Congress Cataloging in Publication Data
Main entry under title:

What? where? when? why?

    (Australasian studies in history and philosophy of science ; 1)
    Includes bibliographies and indexes.
    Contents: Rational expectation and simplicity / F. John Clendinnen –
Why should probability be the guide to life? / D. C. Stove – Chance and degrees
of belief / D. H. Mellor – [etc.]
    1. Science – Philosophy – Addresses, essays, lectures. 2. Science –
History – Addresses, essays, lectures. 3. Induction (Logic) – Addresses,
essays, lectures. 4. Space and time – Addresses, essays, lectures.
5. Salmon, Wesley C. I. Salmon, Wesley C. II. McLaughlin, Robert,
1936–   . III. Series.
Q175.3.W48   113   81–23484
ISBN 978-94-009-7733-4   ISBN 978-94-009-7731-0 (eBook)
DOI 10.1007/978-94-009-7731-0   AACR2

TABLE OF CONTENTS

FOREWORD

Only in fairly recent years has History and Philosophy of Science been recognised — though not always under that name — as a distinct field of scholarly endeavour. Previously, in the Australasian region as elsewhere, those few individuals working within this broad area of inquiry found their base, both intellectually and socially, where they could. In fact, the institutionalisation of History and Philosophy of Science began comparatively early in Australia. An initial lecturing appointment was made at the University of Melbourne immediately after the Second World War, in 1946, and other appointments followed as the subject underwent an expansion during the 1950s and '60s similar to that which took place in other parts of the world. Today there are major Departments at the University of Melbourne, the University of New South Wales and the University of Wollongong, and smaller groups active in many other parts of Australia, and in New Zealand.

"Australasian Studies in History and Philosophy of Science" aims to provide a distinctive publication outlet for Australian and New Zealand scholars working in the general area of history, philosophy and social studies of science. Each volume will comprise a group of essays on a connected theme, edited by an Australian or a New Zealander with special expertise in that particular area. The series should, however, prove of more than merely local interest. Papers will address general issues; parochial topics will be avoided. Furthermore, though in each volume a majority of the contributors will be from Australia or New Zealand, contributions from elsewhere are by no means ruled out. Quite the reverse, in fact — they will be actively encouraged wherever appropriate to the balance of the volume in question. In this as in other respects, this first volume in the series exemplifies what is intended for the series as a whole.

R. W. HOME
General Editor
*Australasian Studies in History
and Philosophy of Science*

ACKNOWLEDGEMENTS

The Trustees of the Estate of Albert Einstein have kindly granted permission for the reprinting of passages from the copyrighted writings of Einstein, appearing here on pages 102, 113, 114, 115, 130 and 253.

Permission is gratefully acknowledged from Fordham University Press to reprint the essay by David Stove entitled 'Why Should Probability be the Guide of Life?'. This essay was originally published in *Hume: A Re-Evaluation*, edited by Donald W. Livingston and James T. King (New York: Fordham University Press, 1976), Copyright © 1976 by Fordham University Press, pp. 50–68.

Warmest thanks are due to Mrs Barbara Young, Philosophy Secretary, Macquarie University, for her endless patience in typing the manuscript of this volume, in its many versions; and for her good-humoured assistance with the myriad administrative details of its evolution.

ROBERT McLAUGHLIN
Editor

EDITORIAL PREFACE

In the Australian spring of 1978, the University of Melbourne enjoyed the exciting presence of Wesley Salmon as a Visiting Professor in the Department of History and Philosophy of Science. In addition to conducting advanced seminars at Melbourne, Salmon spoke at universities in other major Australian cities, including Adelaide and Sydney. Wherever he went, this highly-distinguished American philosopher of science made a telling impact. His work, of course, was already very familiar in Philosophy and HPS circles in Australia; as he was in person to a number of Australians who had been privileged to work with him in the United States. Beyond this, however, his presence here for some three months was stimulating and memorable for the wide variety of people — students as well as teachers, scientists as well as philosophers and historians of science, and scholars generally — who encountered him for the first time. Thus it seems most fitting that the initial volume in this series should take Salmon's work as its *leitmotif*, appreciating his first Australian visit, and exhibiting some of the influence his thought has had upon philosophers of science here, as well as on the international scene.

For three decades, Salmon has been developing progressively a sophisticated empiricist philosophy of science. Yet like his onetime teacher, Hans Reichenbach (who was a Berliner, and neither geographically nor intellectually a member of the Vienna Circle, contrary to popular belief), Salmon would certainly object to being labelled a "logical positivist". Positivism was dualist and instrumentalist — it accorded a privileged status to some reduction base (e.g., physical objects; or (earlier) sense-data) and the language associated therewith, and a derivative or subordinate status to some other entities (e.g., theoretical entities) and their linguistic counterparts (e.g., "constructs"). These marks of an extreme and obsolete empiricism are absent from Salmon's philosophy. For him, as for all empiricists, the sole pathway to knowledge is experience. But this is not to say — and sophisticated empiricists do *not* say — that the only knowable matters (and the only real things) are those accessible to "direct" experience. To say the latter is to commit the fallacy of *epistemologism* (cf. McLaughlin [1967], pp. 15–17) — a fallacy typical of positivism, which Salmon, with other profound empiricists, explicitly avoids.

He is a realist with respect to theoretical entities, and he rejects privileged reduction bases, together with all other forms of apriorism and homocentrism. A deep awareness of the contingency of the world and our knowledge of it — expressed in his denial of natural necessities, occult powers, absolute reference frames, sacrosanct explanatory principles, innate cognitions and synthetic *a priori* judgements — has always informed Salmon's thought. This pervasive sense of contingency (evident at many points in Hume's writings), which is at the heart of his commitment to experience as the only source of knowledge, stamps Salmon as an empiricist — a title which I think he would be glad to wear, even in this mad time (as Clark Glymour might say) when empiricism is under challenge from various quarters, while other, less rational, doctrines are attaining (or regaining) orthodoxy.

Some idea of the scope of his work may be gained from the Salmon Bibliography included in this volume. Without discounting the breadth of his thought, it is fair to say that his major contributions have been in the three areas of induction, space and time, and scientific explanation. When this collection was planned, it seemed natural to organise the material under these headings. The first and last topics are pivotal ones in philosophy of science, at least as it traditionally has been conceived; indeed, Salmon's work has helped in no small way to retain these problem-contexts in their central place. Philosophical problems of space and time also have had an essential, status throughout the history of philosophy — from the time of Zeno, Plato and Euclid to the present day. Current work by philosophers of science in this field can hardly be ignored by anyone who is seriously concerned with contemporary epistemology and metaphysics; while its relevance to the highly active discipline of modern cosmology (via the study of the conceptual foundations of special and general relativity, *inter alia*) needs no emphasis. In this important and difficult area, Salmon has edited one book [1970a] and written another [1975a], as well as many articles.

With respect to the problem of induction, which has occupied him since early in his philosophical career, Salmon has steadfastly refused to take tempting easy ways out. He has always insisted that Hume's problem was a genuine one, not to be repudiated or dissolved by tactics of linguistic analysis, nor to be sidestepped by the deductivist strategy of Popper and his followers. While recognising the inadequacy of Reichenbach's pragmatic vindication of induction, as it stood, Salmon always considered that this was the correct approach to the problem, and has long sought a satisfactory articulation and refinement of it. As he remarks in his 'Further Reflections', his progress along this line met a serious obstacle in a critique offered by Ian Hacking in 1965.

Hacking formulated three criteria which he proved to be necessary and sufficient to single out, from the entire class of asymptotic rules, the one rule
which Salmon sought to justify. Seen in this light, the problem facing Salmon
was to justify these three conditions which are uniquely satisfied by that rule.
At least one of the criteria appeared intractable. Now the opening essay in
this volume, by John Clendinnen, suggests a way forward invoking simplicity
considerations in the form of proscriptions of *arbitrary* choices of rules; and
further enhancements of this new and promising line of development by
Salmon and others may be expected in the near future.

Three other essays extend the discussion of induction and probability.
David Stove takes issue with Salmon, on both historical and epistemological
grounds; after distinguishing a pseudo-problem of induction from the genuine
one, and repudiating the former, he appears to advocate a Carnapian approach
to the latter. In Stove's view, Salmon's (and others') "problem of induction"
is a pathological invention of twentieth-century philosophy, not a product
of Hume's eighteenth-century Age of Reason at all. Hugh Mellor gives an
elaboration of his own version of a propensity theory of probability, which
he believes can avoid the difficulties of frequency theories (such as Salmon's),
as well as defects in other forms of propensity accounts. In the course of this
he vigorously defends himself against a recent critique of propensity theories
published by Salmon. Largely agreeing with Salmon's Bayesian explication
of the logic of hypothesis appraisal, I have tried in my own contribution to
extend his conception of *plausibility considerations* into the context of
invention ("discovery"), as part of an inductivist defence of inventionism.

Within the philosophy of space and time, Salmon's approach to the key
question of the status of simultaneity in special relativity, again perhaps
reflecting Reichenbach's influence, has been rigorously conventionalist. Here
it is worth re-emphasising that the conventionalism of Salmon, Reichenbach
and others (such as Grünbaum) is far removed in fact and in spirit from
positivism, operationism and related dogmas. Instead, it is an expression of an
empiricist rejection of absolutes, privileged reference frames and associated
apriorist conceptions. Salmon's sophisticated empiricist conventionalism is
certainly not to be simply equated with a positivistic insistence on verifica-
tionism, "operational definitions" and similar methodological gimmickry
obsolete for nearly half a century (among philosophers of science, if not
among scientists).

The conventionalist view of simultaneity in the special theory of relativity is subjected to highly original criticism in the two papers dealing with
the philosophy of space and time. John Saunders and John Norton offer a

historico-heuristic argument in their attempt to show that Einstein had "good theoretical reasons" for adopting the value $\epsilon = \frac{1}{2}$, i.e., for taking the one-way speed of light to be the same in all directions. Graham Nerlich, in a provocative and iconoclastic argument couched in Minkowskian spacetime terms, provides an analysis of the concept of a "frame of reference", which he carefully distinguishes from a coordinate system. He proceeds to argue, against Salmon and other conventionalists, that there is no basis in special relativity for assigning a special status to time-like lines, in contrast to space-like lines. His discussion aims to show that the simultaneity of space-like separated events is a matter of relational fact, rather than a convention. Like Saunders/Norton, but using very different arguments, he concludes that the choice of $\epsilon = \frac{1}{2}$ is more than merely conventional.

Salmon's dominant interest over the last few years has been in scientific explanation. His early work emphasised the importance of statistical relevance in explanation, and represented a major departure from the more orthodox version of the "covering law" model, first proposed by Hempel & Oppenheim [1948], and subsequently refined by Hempel in a number of famous essays. Against this model, Salmon has insisted that explanations are not arguments; rather, they are displays of statistical relevance relations, ordinarily expressed probabilistically. A corollary of this approach is that a good explanation need not yield a high probability for the explanandum event(s) — which consequence some critics, such as Mellor [1976], find violently counter-intuitive. In his more recent writings on explanation, Salmon has invoked, and elaborated with great originality, Reichenbach's Principle of the Common Cause, which licenses inferences from statistical relevance relations or correlations to common causes. The resulting account of explanation relies heavily on a conception of continuous causal processes. Salmon's current picture of scientific explanation (which he is still actively developing) is lucidly portrayed in his own original contribution to this volume, "Comets, Pollen and Dreams".

The other three essays treating scientific explanation and causality challenge various features of Salmon's account — most notably his deployment of the Common Cause Principle, and his insistence that continuous causal processes are compatible with indeterminism. Clark Glymour, in addition to offering various "cavils", dignifies Salmon with the title "the last mechanical philosopher". In his "Further Reflections", Salmon gracefully accepts the substance of this appellation, although he doubts that he is "the last" of the species; and he goes on to provide a penetrating analysis of the highly contemporary sense in which he understands the term 'mechanical'. Bas van Fraassen suggests ingenious non-quantum mechanical counterexamples to the

Common Cause Principle, whose application by Salmon in an argument for scientific realism Glymour has also queried. John Forge, after giving a critique of the Hempelian account of explanation within the context of physics, argues that Salmon's approach, too, appears inadequate, in that it cannot accommodate quantum discontinuities (not indeterminism, with which Salmon is confident he can deal). Forge suggests that an important but reparable defect of Salmon's account of explanation is that it is tied to a *particular* theory, namely classical physics, and proposes a way in which it could be generalised to avoid this, while retaining its other features.

The closing element in this collection is Salmon's "Further Reflections". Here, with his usual gentle force, he seeks to answer his critics, and in the course of this to reveal fresh insights of his own. For those interested in Salmon's continuing thought, the "Reflections" constitutes a rich source of his leading themes. It is worth remarking that considerable to-and-fro debate between Salmon and many of the other contributors took place before the volume assumed its final form. For their great patience and good humour in these exchanges I should like to thank all the participants, and notably Salmon himself. One happy result of this dialectic has been the elimination of quite a few misconstruals. Major disagreements remain; but I hope that serious misunderstandings have been minimised.

The distinctive character of this collection is its exhibition and examination of Salmon's current thinking on central problems in the philosophy of science so far unsolved. In this respect it depicts the contemporary status of the questions. Definitive answers to them may well emerge from Salmon's continuing work over the next few years. Whether, in these enterprises, he finally succeeds or not, he already has mightily advanced our understanding of the tasks involved. Recognising his achievement, the present volume essays to yield insight into a crucial stage in the evolution of his thought, and thereby of twentieth-century empiricism.

ROBERT McLAUGHLIN
Sydney
August, 1980

F. JOHN CLENDINNEN

RATIONAL EXPECTATION AND SIMPLICITY

SYNOPSIS. Induction is implicit as a crucial constituent in all scientific inference. A vindication is proposed which argues that only inductively based predictions are rational because they may succeed while any alternatively-based prediction is irrational because it must involve an arbitrary decision and is thus no better than a guess.

Accepting a predicting rule which is more complex than necessary to cover known facts, involves an arbitrary choice. Thus, a simplicity requirement must be written into the specification of induction as vindicated. This meets a number of problems that have beset induction.

0. INTRODUCTION

It may seem that philosophers would always be concerned with the basis of reasoning; yet for a number of decades the status of ampliative inference has been largely avoided. It has been widely held that no justification of non-deductive arguments is possible or necessary. Yet if this so it is hard to see how to avoid the conclusion that all our beliefs about the concrete world lack a rational basis. Few openly proclaim this doctrine, but the position adopted by many implies it.

Wesley Salmon is one of the few philosophers who has faced this issue. His work on induction has been the most serious and sustained attempt to provide a rational justification for ampliative inference. The concern, rigour and thoroughness with which he tackles this task is typical of all his work. Having identified a problem as of major importance, he did not turn lightly away from it. He faced the difficulties which arose squarely and treated his critics generously but, in a long series of articles, showed extraordinary resourcefulness in modifying his justification to meet objections.

Salmon's work is the inspiration and starting point for this essay. I believe that his basic approach is correct but that his assumption that a justification must commence with the convergence of the inductive method was incorrect and that this led to the impasse which he recognised in 1965. (Clendinnen [1977], Salmon [1968b].)

1

Robert McLaughlin (ed.), What? Where? When? Why?, 1–25.
Copyright © 1982 by D. Reidel Publishing Company.

1. INDUCTION AND SCIENTIFIC INFERENCE

What is presented here is intended as a first step towards a justification of the methods which we employ in ampliative inference. What I deal with in this paper is a simplified version of induction. To avoid misinterpretation I hasten to add that I do not hold that much of what goes on in scientific thinking consists in, or is reducible to, simple enumerative induction. However, I do think that induction, explicated in a way that I shall argue for, is basic to scientific inference in general.

The form of induction with which I will deal in this paper is a method of adopting predicting rules. I will use the term "inductive" to characterise any prediction derived from a rule so adopted. It is clear that this is not induction in the usual sense. Nevertheless, accepting a generalisation as true commits one to the soundness of a set of inferences from initial conditions to subsequent facts and thus, in part, consists in adopting a rule for predicting. There is no doubt that many problems stand in the way of dealing with the rationale of theory acceptance. So it seems reasonable to start with a justification of a principle which, while reasonably comprehensive, avoids peripheral complexities.

It is generally accepted that laws and theories are not confirmed merely by reference to their ability to predict and explain particular facts. Nevertheless it is also agreed that this is a very important, and the most directly empirical, part of their evaluation. And in induction the acceptance of rules is based on their past success in prediction. Thus, a justification of this method is a justification of one aspect of scientific method. In this paper I seek a vindication, and in so doing an explication, of induction in the form I have indicated.

2. AN OUTLINE OF THE JUSTIFICATION

Roughly speaking, the inductive method under consideration consists in adopting predicting rules which work in the realm of known facts. This is too vague and must be restated, but it may serve as a start. It is clear enough that we cannot hope to justify this method by showing that it will always or usually work. However, I accept Reichenbach's general idea of a pragmatic justification: that is a justification which establishes that this method is the best way of attempting to achieve the goal of making true predictions.

The vindication which I propose does not involve a proof that induction possesses some special desirable property but is rather an argument that any alternative to induction must possess a most undesirable characteristic, of which induction is free, and which marks the method which has it as irrational. Induction is to be justified by establishing that it may succeed while any alternative method is an irrational way of trying to make correct predictions.

This is essentially a justification in terms of practical reason. The relationship which I believe can be established between the policy of using induction and the end of making correct predictions is a relationship which is generally accepted as justifying a means with respect to a goal when it holds between them. It is universally agreed that when we establish that M may achieve E and that any alternative means is an irrational way of attempting to achieve E then M is the rational way of proceeding when our sole aim is E (and so have a neutral evaluation of M). In this paper I will simply accept arguments of this form as valid, although there is no doubt that the logic of practical arguments needs further attention. (I have made some preliminary remarks in Clendinnen [1977]; Section 5.3.) It should be noted that the argument for the rationality of induction is dependent on, and thus treats as logically prior, the irrationality of alternative ways of proceeding.

In outline the argument is that non-inductively based prediction is, if not itself a mere guess, based on a procedure which includes a purely arbitrary step. As such, non-inductive predictions are no better than guesses. And it is irrational to place any more reliance on one guess than any of the other guesses which might have been made.

In outline the argument is that any non-inductively based prediction is, if not itself a mere guess, based on a procedure which includes a purely arbitrary required to complete a pattern which has always, to our knowledge, persisted then we expect that fact. To expect any other fact would be tantamount to assuming that the pattern amongst future facts will differ from that amongst known facts. But there is an unlimited number of ways in which a pattern may change. How are we to choose amongst these alternatives? Not by reference to the known facts. For how could any set of past facts indicate that some pattern will change in the future? If someone did hold that a certain fact did portend such a change, we would have to enquire how this was known. Why may this fact not be a sign of something quite different? (There is, of course, one legitimate way in which known facts may be taken as indicating that a pattern in the facts will change. This will occur if a specific pattern is seen as falling under a broader pattern and the persistence of the broader pattern implies a change in the narrower one. For instance, the expected diminution in the effectiveness of an antibiotic as resistant strains of bacteria develop. Such cases are, however, instances of induction and not exceptions to it.) Expectations about the future which are not based on facts or are based on them in an arbitrary way are not based on anything but our inclination to believe. As such they deserve no more confidence than any other guess we might have made.

The arbitrary character of non-inductive predictions will be argued for with more attention to detail in the sequel, but first we must consider carefully what is wrong with arbitrarily-based expectations.

3. WHAT IS WRONG WITH GUESSING?

There is, of course, nothing wrong with guesses or arbitrary choices in their proper places. Guessing games are all right and if we want a certain model of a car and there are three unblemished examples, all of the same colour on the showroom floor, our choice will quite reasonably be arbitrary. What is not reasonable is to hold that what is chosen arbitrarily is thereby superior to the options not chosen. Likewise it is unreasonable to hold that any reliance should be put on a mere guess about what will happen.

But first we should consider whether there is sometimes a case for trusting guesses. There are many cases where a person's unreflecting intuitive expectations have been found to be indicators with some degree of reliability. Doctors whose diagnoses go beyond what is possible by standard symptoms and executives who can select prospective employees as suitable on a better than chance basis, are often quite unable to say how they do it. It seems that these are cases of guesses, in an obvious sense; and yet there are inductive grounds for placing some trust in them. Any air of paradox is dispelled when we distinguish the first person from the third person perspective. Anyone acquainted with the doctor's past record has inductive grounds for trusting his diagnosis. And this is true of the doctor himself; for he can view himself objectively. Thus, in placing trust in his own intuitive diagnosis, the doctor is by no means guessing – he is using induction. It is when he considers his subjective experience in forming, as distinct from evaluating, a particular diagnosis, that it seems that he is guessing. And, indeed, there is no way of distinguishing what he does from a guess so long as it is considered in isolation – but in this context there is no basis for relying on his guess.

There is another case where it may seem that arbitrary predictions are rational; however, such cases are misdescribed as predictions. In some situations we must act in one of a number of alternative ways, and we have no basis for deciding which. It may well be that if we knew the outcomes of the alternatives we would have a basis for choice but that we lack such knowledge. It may then seem natural to say that choosing one of the alternatives is acting on the assumption that such and such an outcome will result. It is also clearly reasonable to arbitrarily choose one of the alternatives. We may thus be led to say that it is rational to arbitrarily predict the corresponding outcome. But this is a simple mistake. In such cases, the arbitrary choice of action is rational but to *expect* the outcome we *hope for* is not. (Arbitrarily choosing a horse to bet on so as to have "an interest in the race" is not the same thing as arbitrarily expecting that horse to win.)

3.1. *The Principle of Indifference*

If we agree that it is unreasonable to trust a mere guess we must ask why? For our purpose at least, this question must be answered without appealing to any method of rationally predicting; for the irrationality of guessing is to be a premise in the justification of the most basic method of prediction. How, then, is this question to be answered? I will argue that the guessed outcome of a situation does not differ in any relevant respect from any of the other possible outcomes and that consequently it would be unreasonable to place any more confidence in a guess which is made than in any of the possible guesses which might have been made. A very minimal requirement of reason will then require that we should not expect a mere guess to be realised.

This conclusion might be supported by appeal to the probability calculus and the principle of indifference. When probability is interpreted as degree of belief, the axioms are not mere arithmetical truths; nevertheless, forceful justifying arguments which do not involve induction, have been provided by Ramsey, de Finetti and others. Thus, it would be reasonable to employ this theory in the present enterprise. However, the principle of indifference has a bad reputation. As a basis for numerical probability values, it can lead to paradoxes. For our purpose precise probability values are not required; consequently it may be that a weaker form of the principle which avoids paradoxes may suffice.

We must consider whether the principle of indifference generates paradoxes concerning which alternative is sufficiently probable to accept. Consider the following argument, designed to show that we should expect that an arbitrarily-chosen horse will win a given race. Ten horses, *A, B, C*, etc., are starters. Now consider the following set of possibilities: (1) *A* will win by at least a hundred feet, (2) *A* will win by just ninety-nine feet, (3) *A* will win by ninety-eight feet, (4) . . . etc . . . (100) *A* will win by just one foot (101)

Some other horse will win the race. Now it may be argued that we have no information showing that any one of these possibilities is more probable than any other and that, therefore, the principle of indifference may be applied. We are then led to the conclusion that, in all but one of the hundred and one possibilities, horse A will win. Unless such applications can be blocked, the principle is worthless for even our non-numerical purpose. However, the foregoing paradox depends on an implausible interpretation of the principle – to assume that probabilities are the same whenever we do not know that they are different, is sure to lead to contradictory results. But in its more plausible form this principle takes account of background knowledge about what may be relevant. It applies only to alternatives which are symmetrical in all such respects. In the case before us it is clear that the possibility of horse A winning by just fifty-four feet is symmetrical with horse B winning by just fifty-four feet; and is not symmetrical with the possibility of B winning; let alone with the possibilities of any one of nine horses winning. To expect equally two possibilities which are asymmetrical, means differentiating between one's expectations about possibilities which are symmetrical.

It is central to the very notion of rationally-based expectations that where there is no reason to differentiate the firmness of our expectations, we should not do so. Recognising only the possibility that asymmetries of certain kinds *may* provide a basis for differences in expectations we can formulate a principle in the following terms:

> Given a set of alternative possibilities which are symmetrical in all respects which we have any reason to believe might be relevant to their realisation, and given that we have no positive reason for expecting any one of them to be realised, then it is unreasonable to expect any one of these alternatives with more confidence than any other.

This principle will not always give paradox-free numerical probabilities; for there may be a number of partitions into possibilities which have equal claims to symmetry. This consideration implies that probabilities determined by it must be somewhat indefinite; however, whenever a particular possibility is not distinguished, by anything that we have reason to believe is relevant, from a substantial set of alternatives, it will establish that it would be unreasonable to expect that possibility. And this is all we need.

3.2. *Arbitrary Beliefs*

If I guess that a certain event will occur, there is one thing which clearly differentiates this event from any other possible outcome of the situation; namely that I have guessed it. The question is: is this a relevant difference? I have an inclination to believe, should I follow it? (The question, of course, only arises given the goal of forming expectations which are fulfilled; for if nothing hangs on my beliefs why should I not believe what pleases me?)

On some occasion I may find myself considering different outcomes of the same situation. Even if I do not actually formulate them, reflection will alert me to the variety of possibilities. Granted that I have a capacity to remember my previous thoughts, I will have before my mind a number of possible expectations, and to allow that there is some way of choosing among these is to acknowledge that guesses must be evaluated before being accepted. If, on the other hand, I trust my guess, I proclaim my mental impotence; for doing so implies that thought can provide no guide for beliefs or decisions.

In the context of critically examining the very basis of ampliative inference we cannot *assume* that there are any rational principles to guide our expectations. But equally, we cannot assume that there are none. There may be no point in thought and reflection but in allowing the possibility that these mental activities can properly guide our expectations, we are acknowledging that no weight is to be put on a mere guess; for if we have any grounds for trusting a certain expectation, it is thereby not a guess.

3.3. *Why Think?*

At this point we should, perhaps, ask the most basic question of all: Why should we think? Why not act on the first inclination that comes to mind? Why not simply act? But in asking this very question we are reflecting. We are subjecting our putative course of action to critical assessment. The context of rational assessment is a reflective context.

There is, of course, nothing to prevent us simply abandoning thought. However, acknowledging that we have abandoned thought is not the same thing as holding that it was reasonable to do so. In some contexts it is reasonable to act without reflection, but in deciding that this is so (as we may do in retrospect or in adopting a general policy) we are making a critical, thoughtful, evaluation. The person who gives up thought altogether is not open to persuasion; for his commitment means he will follow no argument. But a person who will think at all may well consider the appropriateness of thought in various contexts. And the limiting question of this kind is whether there is ever any point in thought. Since thought is involved in evaluating any activity, one can hardly challenge the appropriateness of exploring how thought might guide our expectations and actions until the exploration has been conducted — and then only if it has failed. If induction can be justified we will not have failed in this endeavour.

But how do we know that guessing is not as good as thinking? Can we be sure that a particular guess is arbitrary? The subjective awareness of freedom to choose cannot preclude the possibility that the choice is causally determined. This may happen in a way that ensures that the resultant prediction will, in fact, turn out to be true. Thus, it may be argued, a prediction which we take to be a mere guess may actually be objectively determined by factors which also determine the outcome about which we are predicting and may consequently be the best prediction that could be made. This is a genuine possibility but it does not bear on the situation of the person making the prediction. As we have seen above, the situation is different if past successes provide inductive reason to believe that a person's guess will succeed. But someone's recognition that his unguided choice *may* be for the best, and may be causally determined to be so, is no aid to him in choosing. The question is what he is to take into account.

Without getting involved in the philosophy of mind, we can agree that the cases we are concerned with are free acts by agents — 'free' in the sense that they must subjectively make a choice. Whatever unknown causes may influence or determine certain ideas in an agent's mind are irrelevant to his decision about how he will deal with that idea. The causes are irrelevant just because, being unknown to him, he cannot take them into account. He finds himself with an inclination to expect one event but recognises a whole range of alternatives as also possible. He must consider where to place his confidence and what principles should guide this choice.

4. PREDICTING RULES AND FACTS

Can thought rationally guide our expectations? So far we have concluded that if our method of proceeding amounts to guessing, it cannot be rational. This means that reasonably based expectations must be based on facts; for facts are nothing more-nor-less than items of the objective world. If our expectations are not arbitrarty guesses we must employ something objective in forming them. But this is clearly not enough, for we might use facts in a quite arbitrary way. Expectations which are derived from facts in some way which involves a purely arbitrary step, suffer from precisely the same defect as guesses. Once again the expectation is no better than any of a large set of alternatives. Thus, we must consider how facts may be employed in a non-arbitrary way in predicting.

As a first requirement it is essential that predictions be derived from facts according to precisely stateable rules. If a prediction is derived in a way which is free of arbitrary choice it must be possible to specify how this is done. And if it is proper to proceed thus on one occasion it would be an arbitrary choice to do otherwise on any other occasion when the facts to hand were the same in all relevant respects. But predicting from facts according to explicit rules is by no means a guarantee that arbitrary choice has been excluded; for the rules themselves may well be arbitrary. Indeed, a little ingenuity would always enable us, in any given situation, to formulate a rule which would apply only to that situation and predict whatever we wished.

Not only must we predict from known facts but we must do so by using rules which are themselves based on known facts in a non-arbitrary way. The adoption of any other kind of rule would involve an arbitrary choice which would ensure that the resultant predictions were no better than any of the alternatives which would have resulted if a different rule had been used in making the prediction. On any particular occasion, the only facts we can use are those which are known to us. Thus, the question is: is there any one non-arbitrary way of basing predicting rules on a finite set of facts? Undoubtedly there are an infinite number of relationships which might hold between a rule and a set of facts, but a non-arbitrary method must employ a direct relationship; that is a relationship which does not involve any entity other than the rule and the set of facts. Suppose, for instance, that an emperor proclaims that, if facts conforming to a certain pattern occur, then a certain predicting rule is to be employed. This immediately raises the question: why accept the emperor as an authority in this matter? There may be competing "authorities" and there are certainly many alternative "methods" he may have proclaimed. It would be arbitrary to accept this method unless there is some reason for so doing. Some facts might be cited for trusting the emperor on this matter — maybe he has been recommended to us by some highly revered prophet. But then we must ask: "Why trust the prophet in this matter?" We are confronted by a regress. At any stage, facts might be cited as a justification for accepting a rule or method, but in each case, we must either accept without question that the facts are relevant in the way they are held to be or we must seek reasons. Unless the facts relate to the method directly, the first course must be arbitrary; and in following it we would ensure that our consequential predictions were no better than guesses. Whereas the second course will simply extend the procedure to another stage where precisely the same question arises. This regress can only be concluded by an arbitrary choice.

There are many non-inductive ways of predicting which are, on the face of it, not mere guesses. There are various ways of fortune telling which employ facts and proceed according to definite rules. Now it is, of course, possible that any one of these methods might acquire a history of success. In this case, following the method would no longer be a non-inductive way of predicting and, as I will argue, inductive methods are not arbitrary. However, to adopt such a method with no evidence of its past success, is an arbitrary choice which might just as well have been made differently by adopting an alternative method. And if some reason, other than past success, is given, the appeal to a consideration of that kind will itself call for a justification if it is not directly arbitrary.

If we are seeking a direct relationship between a rule and a set of facts, we must consider whether there is any relationship involved in the very concept of a rule. Now it would be impossible to comprehend what a predicting rule is without understanding what it is for a rule to succeed or fail in an application. The latter notion is implicit in the former. It is equally clear that facts are involved in such successes or failures. It is clear that given a rule and a set of facts, then just one of the following relationships must hold:

(a) The rule may be inconsistent with the facts because the predictions derived from it and some of the facts are falsified by other facts.

(b) The rule may be vacuously consistent with the facts because it does not facilitate any predictions which are verifiable by any of the facts in the set.

(c) The rule is consistent with the facts and exemplified in them in the sense that some of the facts with the rule entail predictions which are confirmed by other facts in the set.

Relationships (a) and (b), if used for selecting predicting rules, would either give many rules which would be incompatible with each other (in the sense that they would lead to contradictory predictions) or involve arbitrary choice; depending on whether the method directed that all rules in that relationship to the facts were to be adopted or allowed that any such rule may be adopted.

A number of rules may all be consistent with, and exemplified by, a set of facts; but in this case we may expect that these rules are all consistent with each other (just because they do all apply to the same set of facts). Thus it seems we have here one relationship which will select rules consistently and is based on the concepts of application and success. And these latter notions are central to the very idea of a predicting rule. This establishes a *prima facie* case that the method employing this relationship is the one and only direct way of using facts as a basis for adopting rules. The argument is one of conceptual analysis; however, there is no simple analytic truth which establishes the unique status of this relationship. The force of the argument must depend on the unavailability of a counter argument that there is some other comparable relationship.

It has been suggested that there is a straightforward alternative to induction which is simply its opposite. Were this correct, it might be argued that if induction is directly based on facts, then its opposite must equally be. However, there is no consistent general method of making specific predictions which is the opposite of induction. This can be seen when we remember that in using induction we act on the hypothesis that the patterns present in known facts will persist. If we adopt any alternative to induction, we are assuming that at least one pattern will change in some way. The opposite of induction presumably assumes that each pattern will be replaced by its opposite. But there is no specific pattern which is the opposite of any given pattern. In simply denying that a given pattern will persist, we leave open a whole range of possibilities. Consider, for instance, the regularity of litmus always turning red in an acid solution. What is its opposite? The mere denial that this is so is compatible with any finite set of events and so facilitates no specific predictions. The claim that in acid solutions litmus always turns non-red is better but still extremely imprecise. Does it always remain blue? Does it sometimes turn yellow? Or perhaps always? Or perhaps some other colour? The notion of counter-induction is simply not a viable method of predicting. As Salmon has shown ([1956b]), it is liable to lead to contradictions unless hedged around by constraints which ensure that it is incapable of making any kind of precise predictions.

Which Facts are Relevant?

Given that we will adopt rules which are compatible with and exemplified amongst known facts, just which of the known facts will we employ? Before answering this question we must note that there will be criteria of relevance which can be recognised for any particular rule which is under consideration. For example, suppose we wish to assess the rule "Predict that anything which is *A* is *B*". As far as its conformity to facts is concerned, it will be unnecessary to further investigate any individual which is already known to be not-*A* or *B*; whereas those which are as yet only known to be *A* or not-*B*, must be further investigated. Thus, there are some facts in the realm of the known which may be ignored in checking a particular rule. But beyond this are we entitled to ignore other facts? May we select some subset of known facts and restrict our investigation to these, irrespective of the forementioned relevance consideration?

The consideration guiding our choice of method is that it should be free of arbitrary decisions; consequently we may not arbitrarily neglect any set of facts. Neglecting certain facts could make a substantial difference to the rules we adopted. Some rules may come to be rejected through being merely vacuously consistent with the limited set of facts examined while being exemplified in those ignored. Other rules may be accepted, even although they were falsified by some accessible but ignored facts. Just which facts are ignored will determine which rules are dropped or which additional rules are accepted. Thus an arbitrary decision to ignore some facts, will lead to the arbitrary omission or adoption of predicting rules and, hence, to arbitrary predictions.

It may be suggested that it is just as arbitrary to be guided by the whole set of known facts as by any proper subset of it. However, we are committed to employing known facts and this implies that unless there is some reason to the contrary we should take account of any fact which we know could influence our conclusion. To arbitrarily select and use some of the available facts would be a very strange way of being guided by the facts.

5. NEEDLESS COMPLEXITY

We must now turn to a problem which may seem to vitiate the justification developed so far; and this will lead us to an important modification of the usual specification of induction. If there is no constraint on what is to count as a predicting rule, we may use the following recipe for predicting whatever we wish by a method which would, on the foregoing account, seem to be induction. Suppose we have observed that all known things which are *A* are also *X*. In the face of this we may predict that the next *A* will be *Y*, where *Y* is incompatible with *X*, and do so by adopting an apparently inductive rule. Suppose the occasion on which we are making the prediction is at time *T*. Now consider the rule:

(1)   Always predict that an A is X up to time T and thereafter predict that an A is Y.

It is equally consistent with known facts and, indeed, would have made precisely the same predictions as the more obvious rule:

(2)   Always predict that an A is X.

Thus the method of adopting rules which are consistent with and exemplified amongst the known facts, would licence us to adopt the first just as much as the second rule; and also an infinite set of alternatives modelled on the first. But all of these rules differ with respect to some, at least, of their future predictions. The method which directs us to adopt all of these rules is inconsistent and would, therefore, fail to identify any specific expectations for it produces all possible predictions about the future. If we adopt the method which merely allows us to adopt any such rule, then this method effectively allows a purely arbitrary choice determining which of all possibilities we expect.

Such queer predicting rules are obviously reminiscent of Nelson Goodman's paradox of induction ([1965]; p. 72 *et seq.*); but it is important to note the difference. These predicting rules use only English and do not depend on the special predicates devised by Goodman. The force of his paradox was to demonstrate that we cannot hope to explicate induction in purely syntactical terms; for given generalisations which are syntactically complex in certain respects, it is always possible to define predicates which can be used to restate the original generalisation in the same form as the simplest of generalisations. For example, the generalisation "All emeralds are green until 1990 and blue thereafter" can be restated as "All emeralds are grue" given the appropriate definition of "grue". (Namely "x is grue at t if and only if x is green and t is prior to the end of 1990 or x is blue and t is after this date". This definition differs from Goodman's own definition, but serves to make his point and has been adopted by a number of authors who have written on this issue.) Now any criterion adopted in the explication of induction, which excluded the original generalisation on purely syntactical grounds would be circumvented by the use of the queer predicate "grue".

Thus, Goodman points out, we must employ some criterion which controls the use of predicates in induction. However, there is a complementary point on which Goodman does not dwell but which must not be neglected. A satisfactory explication of induction must also provide criteria for excluding queer predicting rules which use only English predicates but at the expense of syntactical complexity. We must consider what requirement will achieve this and low this requirement may be justified.

If we consider the rule

(1)   Always predict that an A is X up to time T and thereafter predict that an A is Y

we see that it is a conjunction and that the second conjunct is only vacuously consistent with known facts: no things which are A existing after T have been observed and so the second conjunct of the rule is not exemplified. We have seen that the policy of adopting rules, which are not exemplified amongst known facts, involves arbitrary choices and is therefore to be rejected. It would be quite unreasonable to exclude such rules when they stand on their own but admit them when conjoined with other rules which are exemplified; for a rule admitted in the second way would lead to precisely the same predictions as if it had been admitted in the first way. Thus, to avoid arbitrary predictions, we may require that a conjunction of rules is adopted only when each of the conjuncts separately meets the requirements for adoption.

This requirement will exclude the kind of queer predicting rules we are considering when they are expressed in the form of a conjunction, but it is also necessary to meet Goodman's point that any such rule can be stated in the syntactical form of a simple sentence, involving no conjunction, by introducing predicates like "grue". Goodman believes that the only effective criterion for excluding these predicates is the fact that they have not been used in actual induction in the way that "green" and "blue" have been; that is, they are not entrenched. This criterion depends entirely on the way people have used words. Goodman says

the judgement of projectability has derived from the habitual projection, rather than the habitual projection from the judgement of projectability. The reason why only the right predicates happen so luckily to have become well entrenched is just that the well entrenched predicates have thereby become the right ones. ([1965], p. 98).

Thus on his account it is not the case that there is something unsatisfactory about "grue" which has resulted in its not being used in past inductions (which would have all been verified in experience to date). Rather, the reason for not using it now is simply that it has not been used; and if it had been, it would have been proper to expect that emeralds will remain grue. Thus for Goodman the propriety of our predictions is entirely at the mercy of how people happen to have used words in the past. Our current expectations are to be dictated by the tradition of expectations which has persisted in our society without there being any possibility of establishing the rationality of that tradition. I find this position utterly unsatisfactory. If this were, indeed, all that a prediction being rational amounted to, I do not see how rationality would recommend itself to us. Some people are, of course, natural conservatives, and they would use entrenched predicates; others are more independent and would sometimes not. If Goodman were right, there would be no more to be said. Why conform just for conformity's sake?

If we look for some reason other than entrenchment for excluding inductions based on Goodman's queer predicates, a plausible candidate is an extension of the conjunction requirement referred to above. It may be suggested that no rule which is logically equivalent to a conjunction is inductively supported, unless each conjunct is. However, this proposal is disastrous. Goodman has pointed out that any simple generalisation (and the same point holds for rules) is logically equivalent to a conjunction employing queer predicates like "grue". For example, if "bleen" is defined similarly to "grue", but with "green" and "blue" changing places, then it is possible to equate

(3)   Always predict that an emerald is green.

with

(4)   Always predict that an emerald is grue at any time up to T and thereafter predict that any emerald is bleen.

The extended conjunction requirement tells us that the second rule may only be adopted if the rule

(4a)   predict that any emerald at a time after T is bleen.

is inductively supported. But this rule is not exemplified amongst known facts and is therefore not supported. The extended conjunction requirement would mean that the simple rule about emeralds cannot be adopted, for it is logically equivalent to a conjunction, one conjunct of which is not supported by evidence. This requirement would exclude too much; indeed it would rule out the possibility of induction without a hearing.

It is possible to express the basic idea of the simple conjunction requirement in a way which does not refer to syntactical form and which thereby gives it a broader scope. By adding a conjunct, which is vacuously consistent with known facts, to a conjunct, which is exemplified, we get a rule with a component which is quite unnecessary to achieve what is required of an acceptable rule. Thus, the known facts play no part in the adoption of this component. Yet in the cases we are concerned with, the component does determine some of the predictions made by the rule. Thus these predictions are based on an arbitrary choice. So in seeking a non-arbitrary method, we must reject the inclusion of any such component in a rule. We have also seen that precisely the same end may be achieved without adding any syntactically-distinct component to the rule but rather by employing a suitable predicate. In the first case a complexity which is unnecessary to fit the rule to known facts is added in the form of a separate clause; in the second case, the same needless complexity is built into the predicates employed. In either case, the need to exclude such needless complexity follows from the policy of developing a method free of arbitrary decision.

To make a predicting rule less simple than is necessary to ensure that it is consistent with and exemplified in known facts, and in a way which makes a difference to the predictions made, is to introduce a purely arbitrary element into the rule and, hence, into the predictions derived from it. Using such rules in predictions is to predict in a way which is not guided by known facts; for it depends on decisions which are no more than guesses. There are an unlimited number of ways in which a predicting rule may be made more complex and there are no constraints on the degree of such complexity. For example, we may vary the rule "Predict that emeralds are green until T and blue thereafter" by giving any value to T or by changing "blue" to any other colour. Or we may add any number of further dates at which similar changes in our predictions are to be made. The choice among all these alternatives must be purely arbitrary for known facts only influence that part of the rule which covers predictions up to the present. The same thing is true if equivalent rules are formulated in a syntactically simple form by using suitably complex predicates. The arbitrary aspect of all such rules can only be eliminated by adopting the simplest rule which fits known facts. In this way, the facts determine the rule and the person adopting it is not left with unguided choices.

5.1. *The Complexity of Grue*

I have argued that Goodman's queer predicates are less simple than the corresponding ones of natural language, on the grounds that their use leads to rules which are equivalent to rules which are complex in an obvious sense. However, Goodman has, in effect, argued against this. He argues that his predicates are no more positional with respect to time than are the English predicates they replace (p. 78 *et seq*.). His point is that it is possible to define "green" and "blue" in terms of "grue" and "bleen" and in so doing there would be the same reference to time T as there is in the definition of "grue" and "bleen" in terms of our familiar predicates. It is suggested that this shows that neither pair of predicates is either more or less positional with respect to time, than is the other. However, this argument is invalid for it overlooks the possibility that the use of a time reference in defining "green" and "blue" merely serves to cancel out the temporal positionality which exists in "grue" and "bleen". To suggest that both sets of definitions stand on an equal footing, is to forget that a definition only gives meaning when the terms used in it already have meaning. Goodman assumes that any asymmetry between the two pairs of terms must be expressed in the definitional relationships between them. However, we must remember that "grue" and "bleen" only gain meaning by being defined in terms of predicates which have an established place in our language and which are in no way temporally positional. What the symmetry between the two pairs of predicates shows is that *if* "grue" and "bleen" had been non-positional predicates and "green" and "blue" had been introduced by definitions that retained the relationship, *then* the latter predicates would have been positional.

That "green" is not time-positional is shown by considering what is involved in learning to use this predicate and comparing this with what would be necessary to learn the word "grue". In learning any colour word in English we must learn to recognise a similarity and continuity among various surfaces with respect to aspects of their visual appearances and be able to recognise which new surfaces fall within the range of similarity. This is involved in learning both "green" and "blue". To use "grue" correctly, persons must be able to recognise one range of colour similarity up to T and a different one thereafter. They must also know when to employ the one standard and when the other. They must learn this whether or not they learn English or other words which designate the most direct colour concepts. Only by going through this procedure could anyone learn to use the word "grue" in the way that results from following the definition.

Thus, it is clear that "grue" is time-positional and thereby more complex than "green", which is not. A predicting rule which used "grue" when the use of the simpler concept would be sufficient to ensure conformity to known facts, is thereby needlessly complex and will be excluded by a principle of induction which includes a simplicity requirement. Some will hold that my account of "grue" is erroneous. In oral discussions of this ingenious and provocative predicate I have found some people who hold that a person could learn to use "grue" in precisely the same way that "green" is learnt. But this must be wrong. Suppose someone was presented with a number of examples of things which were grue and succeeded in recognising the appropriate way in which they were perceptually similar, so that he could correctly apply the word to other objects. It may seem that he has learnt how to use the word. But if this is the full extent of his instruction, what he has learnt is to use the word "grue" in just the way that we use the word "green". He has not learnt a word with a new meaning; and the whole point of Goodman's example is that the use of "grue" in inductions leads to expectations different from those resulting from the use of "green" and, thus, has a different meaning.

5.2. *Simplicity and the Subjectivity of Experience*

The simplicity requirement which I am proposing applies to predicates, and the simplicity of a predicate is to be judged by considering how directly it relates to our experience. But, it may be asked, why should we assume that the subjective character of our perceptions reflects the structure of the external reality which gives rise to them? A property which is experienced by us as simple may be caused by an object of complex structure. Certainly predicates like "grue" have no place in scientific theory but the predicates which are basic in theory are typically far removed from those we employ in describing our experience. Should this not warn us against the policy proposed?

It should be clear that what is proposed is a methodological principle of simplicity which must be contrasted with what some authors have called a metaphysical principle (see Schlesinger [1963] for this distinction). That is, the principle does not claim that correct predicting rules are simple but merely directs us to avoid rules which are less simple than necessary. This may mean that as we accumulate more and more detailed data our rules become quite complex; and it may be that this involves the introduction of predicates which are far removed from our immediate experience. We must certainly acknowledge that the qualitative distinctions we are immediately aware of may be quite inadequate to express the theories which are the simplest basis for detailed predictions; but nevertheless, the only basis for adopting such theories, and the abstract predicates they employ, is the complexity of data gained in immediate experience. The principle that we should only move on from predicates based directly on experience when we have reason to do so by no means denies that such a move may be necessary.

6. OVERALL SIMPLICITY

I have argued that the only non-arbitrary method consists in adopting predicting rules exemplified in and consistent with the set of known facts, provided that each rule is the simplest needed to cover those facts it does cover. What this principle does not do is give us any idea of how wide a range of facts any particular rule should cover. The requirement that we must neglect no known facts, does not help here.

There are many cases where this problem becomes trivial. Consider one rule which predicts the location of the appendix in blue-eyed men and another rule which makes the same prediction for brown-eyed men. It is obviously more convenient to simply adopt a single rule dealing with men in general; but nothing at all hangs on this – the same predictions will be made in either case. We might invoke simplicity in favour of the more general rule. But this would be the notion of descriptive simplicity: a principle of mere convenience which chooses between logically equivalent alternatives. The principle of simplicity invoked in the last section is quite different. It deals with what Reichenbach ([1961], p. 373) characterised "inductive simplicity"; for it plays a positive role in the expectations which we form.

The history of science shows us how often a single generalisation can cover phenomena which differ markedly at the level of visual description. For instance, the law about the expansion of bodies when their temperature is increased, although of a low level of theoretical abstraction, covers phenomena which look very different – compare the apparatus for use in a teaching laboratory where steam is passed through a copper tube with that for demonstrating the expansion of a gas at constant pressure. Such cases are of more practical significance than our initial hypothetical example but it again seems that the single broad predicting rule is equivalent to the conjunction of the narrower ones which jointly covered the same range of phenomena. Indeed, it may be suggested that nothing more than the principle of descriptive simplicity is involved in all cases where we make decisions about the scope of a law.

Such an argument would assume that we always deal with fully determinable facts, but this is a quite unreal idealisation of the actual situation. Because of the smallness or distance of objects, it is often only possible to observe their grosser features; quantitative properties can only be determined to a limited degree of exactness; we know about many events only indirectly by inference and there are many entities which we have good reason to believe exist but which are, of their very nature, unobservable. This feature of knowledge determines the importance of theory. It leads us to employ abstract and idealised properties and to postulate the existence of micro-entities and events remote in space and time.

This feature of knowledge, and scientific knowledge in particular, also means that the predictions we are led to make are often dependent on the scope of phenomena which are dealt with together. For instance, Newton's theory of universal gravitation only emerged when he treated together the motion of bodies near the earth's surface, the moon and planets; and from this theory predictions, for instance, about the existence of a previously unsuspected planet, resulted which otherwise would not have been made. It seems natural to seek generalisations which are as broad as possible and to prefer theories which achieve this; but can this policy be justified?

We have seen that arbitrary choice would be involved in adopting a single predicting rule which was more complex than it need be. This can be generalised to apply to our whole system of laws and theory. If the character of this system is to be determined by known facts, it must not depend on arbitrary choice. To allow that we may adopt a system which included complexities not needed to fit it to the body of directly-known facts would allow for purely arbitrary choice; for there is no limit to the number of ways in which a theory may be made more complex while still fitting the known facts. Thus, our decisions about overall theory structure will only be guided by the facts if we agree to adopt that systematisation of known facts which is simplest.

This principle will provide a guide concerning the scope of any particular rule. It may be that a law or theory must be made more complex in order to increase its scope. This will be acceptable if the result is greater simplicity of the whole system. It is reasonable to make one law rather more complex if this renders unnecessary a number of laws which were previously required to cover the phenomena in the extension to the scope of the first law.

It is now possible to state the principle of induction in a finally acceptable form:

> Adopt the simplest system of predicting rules which are compatible with, and exemplified in, the set of known facts.

In dealing with a particular rule, this means that it is kept as simple as possible with the proviso that increased complexity may be acceptable if this extends its scope. This will depend on an evaluation of overall simplicity.

Such decisions would only be strictly univocal if there were a single uncontroversial measure of simplicity which covered all its different aspects. It cannot, however, be claimed that such a measure is agreed on or is implicit in our concept of simplicity. This does not mean that there is ambiguity in every comparison of simplicity. If two rules differ by the second including all that is in the first and additional components as well, there is no doubt that the former is simpler. In general, if the number of components of a certain kind is less in one rule than the number of components of that same kind in a second, and they do not differ in other respects, then the first rule is simpler than the second. This criterion is implicit in our concept of simplicity and, as we have seen above, is sufficient to settle some important questions about the application of induction. What we cannot be sure of is that there may not be occasions in which one system of rules is simpler than another in certain respects, but more complex in other respects, and where different people disagree about how these balance out. However, this feature of the method explicated here, does not tell against it being the core of scientific method; for the history of science assures us that there are, often enough, disagreements amongst people of competence about which of competing theories is best supported. The method which I claim has been vindicated above, does not give us an algorithm for adopting predicting rules but it does provide a criterion which will, in many cases, be decisive; and where it does not, it does determine that consideration which should guide our judgement in making choices.

A General Methodology

It is widely agreed that the adoption of scientific laws and theories is not determined solely by their conformity with the facts given in experience. I would claim that the requirement that our whole system of science must be as simple as possible is all that needs to be added. Where new facts are more or less unrelated to already-accepted theory, this principle determines the form of the laws adopted. In other cases it serves as a guide in integrating new information into our already-accepted body of knowledge. From this perspective we can understand how apparently different criteria come to be employed in the interpretation of data at different times in the history of science. It is not that our standards of rationality change but the context in which they are applied does. In each period scientists seek the simplest overall system of theory; but as the background of accepted theory changes, so do the kinds of laws and specific theory which can most readily be integrated. It is also possible to see how a major revolution in theory can take place; for at any time it may be recognised that considerable simplification can be achieved by adopting a new conceptual system.

The extension of the inductive principle as a general methodology of science is a quite complex matter. It will involve its adaptation to inferences to theories and existential hypotheses. It will also be necessary to consider how we may evaluate degrees of inductive support and deal with conflicts between the desideratum of having rules which cover as wide a range as possible and the need to adopt only reasonably well-supported hypotheses. I believe that the principle of overall simplicity, and probability theory, via Bayes' theorem, can solve these problems; the former principle providing a basis for comparisons of initial probabilities. However, there is not space here to explore these matters. I hope to do so elsewhere.

7. CROSSING SPATIO-TEMPORAL GAPS

In conclusion, I want to deal with an objection which has been raised in discussion and which touches on the very nature of what is done in induction or any rational attempt to predict.

This objection could be expressed simply if the "conjunction requirement", discussed in Section 5, were accepted. Any predicting rule is equivalent at any given time to the conjunction of two rules, one dealing with the past and one dealing with the future. Now this requirement says that we may only accept the original rule if each of the component rules is inductively supported. But the second rule can have no such support and the first rule can give no predictions. Thus, it might seem that inductive predictions are not possible.

We saw, however, that the extended conjunction requirement has counter-intuitive implications and lacks plausible support. We might reply more specifically to the argument in the previous paragraph by remarking that, since the general predicting rule does have inductive support (and this, of course, presupposes the justification of induction), the predicting rule dealing with the future is *thereby* also supported (for the former entails the latter). The argument would only be valid on the assumption that the only way of supporting the general rule was by supporting both conjuncts.

An argument on similar lines, but with less emphasis on the language involved, might be mounted as follows. In formulating a rule about a set of entities, it is appropriate to consider whether it applies to various proper subsets. If we know of instances where it does not apply to some, or all, members of a certain subset then the original rule is certainly falsified; but a more restricted rule might still be adopted. If some subsets have not been examined to determine whether or not the rule applies in their case, this constitutes a weakness in the induction. Now, it may be argued, there is always one subset of entities in which no rule has ever been tested; namely the set of future events. Hence, it seems that we always lack relevant data about the very set of entities about which we wish to form expectations.

The question that is raised here concerns the kinds of sets over which we may legitimately generalise in induction. To require that inductive evidence must be available about every subset of the set covered by a rule, would be to require that every single individual be examined (for the set of subsets includes a singleton set for every element of the original set). This would be tantamount to holding that we can only form rational beliefs about any event by directly knowing that event; that amounts to denying the validity of any ampliative inference. Nevertheless, requiring that evidence about both subsets is necessary when a significant partition can be drawn would seem to be a very reasonable requirement. And there is surely a case for holding that the partition of events into past and future is significant.

Consider the following argument. If there were two sets of entities which, while having some properties in common, differed in some obvious respect, we would not think of formulating a generalisation covering the union of these two sets, unless members from both sets had been examined to ensure that it held for them. We would not think that one set of things was a reliable indicator about another set whose members obviously differed from those of the first set. Now why do we think that the perceptual properties of events are more significant than their location in time? Surely the division into past and future is of major significance for any epistemological purpose. It is true that we do not think of the temporal, or spatial, differences as significant in our classification of events and things into those categories which we think of as "natural". But this is surely an accident of our psychology; yet on it depends the apparent legitimacy of the whole inductive method. So goes the argument we must face.

On reflection it can be seen that we are confronted by an argument based on the assumption that induction has not been justified; rather than one which challenges the justification which has been proposed. If this, or any other, justification stands then we may rely on this and not our psychological inclinations in inductive projections from what is observed to what is not. The fact that we are predisposed to think in this way is no more of an invalidation than it is a validation of so doing. There would only be a criticism of a method if it left no alternative but to trust these predispositions.

The starting point of the justification of induction is our concern with the part of reality with which we are as yet unacquainted: the future and those things outside the spatial region we have explored. The question then is: can we form reasonable expectations about what has not yet been experienced? And I have argued that just one method of forming expectations is reasonable for the modest reason that it may succeed and any alternative way of proceeding would be irrational. The whole perspective of the problem and the proposed answer involves a dichotomy between spatio-temporal locations on the one hand and, on the other hand, the whole range of qualitative and quantitative similarities and dissimilarities between things and events. When we use information about what is known directly in forming expectations about what is not, differences of the former kind mark off what is referred to in the premise from that in the conclusion. Classifications of the second kind, however, are the basis of the predicting rules we use. The possibility of formulating such rules, or indeed, of making predictions depends on our ability to detect these similarities and differences; for it is only in this way that we can relate things which are separated in space and time. It would necessarily defeat the whole project to employ the former kind of property for the latter purpose and suggest that sameness of spatio-temporal location be used in determining what counts as inductive evidence for what; for to do so would be to reject the notion of inductive evidence and insist that we can only rationally form beliefs about individual events by direct acquaintance. As long as the rationality of induction is an open question we must allow the possibility of crossing spatio-temporal gaps.

8. WHAT HAS BEEN ACHIEVED?

Many will feel that the vindication offered here establishes too little to be of any value. People ask "Can we trust the predictions of science?" In reply they are told: "These predictions may succeed and it would be irrational to accept any others". One might be excused for feeling that the response is just too weak.

We must remember, however, that the task was to establish that induction is a reasonable way of proceeding. Given this as established, it is possible to recognise that in the present context of knowledge there is vastly more than this bare claim to be said for the inductive method. P. F. Strawson ([1952], p. 261) stresses the importance of distinguishing two distinct propositions:

(I)   (the Universe is such that) induction will continue to be successful and

    (II)   Induction is rational.

He points out that given grounds for accepting (II) there is no inconsistency at all in accumulating inductive evidence for (I). I disagree with what he says about the basis for accepting (II), but I believe that what he says about the relationship of (I) and (II) is of the utmost importance to a correct perspective on the justification of induction. It points to the element of truth in the uncritical but appealing idea that induction is justified by its past success.

There must be a non-inductive justification of induction; but in establishing it as a reasonable policy, nothing is settled about what regularities there are to project. The extent of regularity in the world we have experienced to date is attested by the extraordinary achievements of science. The significance of these past achievements for the future has been my concern in this paper.

University of Melbourne

D. C. STOVE

WHY SHOULD PROBABILITY BE THE GUIDE OF LIFE? [1]

I

(Q1)   Why should one believe a proposition H which is certain in rela-
         tion to one's total evidence E?

Because (A1), necessarily, if one's total evidence E is true, then every
    proposition H which is certain in relation to E is true.

Let us assume that (A1) is an appropriate and adequate answer to the question (Q1). I do not think it is, and it seems obvious to me that there is even
something seriously wrong with the question itself. But many philosophers
think otherwise, and I intend to proceed on their assumption for the first
half of this essay.

Now consider the question, analogous to (Q1):

(Q2)   Why should one believe a proposition H which is probable but
         not certain in relation to one's total evidence E (since any such H
         may be false)?

Is there an adequate answer to (Q2) which is analogous to (A1)? One possible
answer to (Q2) is:

Because (A2), necessarily, if one's total evidence E is true, then every
    proposition H which in relation to E has probability $= x/y$ (where
    $x/y > 0 < 1$, but may be close to 1) has probability $= x/y$ in
    relation to E.

What (A2) says is true, but it does not seem an adequate answer to (Q2)
and perhaps not even an appropriate one. It is, rather, a repudiation of that
question. The proviso it contains, that E be true, is clearly redundant. And
once that is omitted it is evident, if it was not so before, that (A2) simply
says that one should believe what is probable in relation to one's total evidence
because it is probable in relation to one's total evidence. But that seems too
short a way with the dissenter who asks (Q2). I do not deny that there is
something seriously wrong with (Q2): on the contrary, it seems obvious to
me that there is. But, then, there is also something seriously wrong with

27

Robert McLaughlin (ed.), What? Where? When? Why?, 27–48.
Copyright © 1976 by Fordham University Press, New York.

"Why should one respect what is sacred?" — yet that question can have merit. It does not always deserve to be repudiated with "One should respect what is sacred because it is sacred". But as an answer to (Q2), (A2) seems to be no better than that.

A second possible answer to (Q2) is:

> Because (A2′), necessarily, if one's total evidence E is true, then every proposition H which in relation to E has probability = x/y (where $x/y > 0 < 1$, but may be close to 1) has (simple) probability = x/y.

This is not an adequate answer. The conception of simple probability — that is, of the probability of a proposition in itself, as contrasted with the conception of the probability of a proposition in relation to another proposition — is a mysterious one in itself, and to give this answer to (Q2) would be to invite in its place, another question which would prove harder still to answer, viz:

> Why should one believe a proposition H which is (simply) probable but not certain (since any such H may be false)?

But in any case, it is disastrous to *connect* the simple or one-placed conception of probability with the two-placed conception, in the way in which (A2′) does, as the following well-known argument shows.

"Tex is rich" has probability = 0.9 in relation to "$\frac{9}{10}$ths of Texans are rich and Tex is a Texan"; and it has probability = 0.1 in relation to "$\frac{1}{10}$th of Texan philosophers are rich and Tex is a Texan philosopher". The former conjunction might be my total evidence, the latter yours. Both could be true. If they are, then given (A2′), it would follow that "Tex is rich" has (simple) probability both = 0.9 and = 0.1.

A third possible answer to (Q2) is:

> Because (A2″), necessarily, if one's total evidence E is true, then a proportion = x/y of the propositions H which each have probability = x/y in relation to E are true.

This seems an appropriate answer, and perhaps would even be adequate if true. But it is not true. "There are just 100 tickets in Lottery L_1, and L_1, is fair, and I hold just one ticket in L_1" might be true and might be my total evidence E. The only proposition H which in relation to this E has probability = $\frac{1}{100}$ is "I win L_1". But it is not only not necessary, it is impossible, that in this unit-class of propositions a proportion = $\frac{1}{100}$ should be true.

The case is no different where the number of propositions H each having probability $= x/y$ in relation to E is such as to allow the possibility of a proportion x/y of them to be true. My total evidence E might be true, and might be such that the only propositions H each having probability $= x/y >$ $0 < 1$ in relation to it were the hundred hypotheses "I win Lottery L_1", "I win Lottery L_2" and so on, to "I win Lottery L_{100}". Common sense says "You can't win them all", and so says a certain immemorial distortion of classical probability-theory; and so says (A2''). But of course I *can* win them all, or none, or any other proportion, and what (A2'') says is false. This is really excessively obvious and is, of course, no more than almost every reputable writer on probability has always allowed. At the same time, it is necessary to insist on it, because there is a permanent temptation to distort the Law of Large Numbers so as to make it testify to the contrary.

That there is no adequate answer to (Q2) does not, of course, follow from the fact that the three answers I have now considered are inadequate. Nevertheless, this alarming conclusion is already looming up, because there is, in fact, in the answers to (Q2) already found wanting, at least an approach to exhaustiveness. To explain.

What is wanted is an answer to (Q2) which differs from (A1) in some respects (viz., those required by its being an answer to a question about probability rather than about certainty), but which also preserves some features of (A1) intact.

One feature of (A1) which any answer to (Q2) must preserve is the begin-ning "(Because) *necessarily* ... " To begin an answer to (Q2) with "(Because) it is probable though not certain in relation to our total evidence that ... " would be inadequate for an obvious reason. To begin it with "(Because) it is (simply) probable that ... " would be inadequate for other obvious reasons; among them, the inviting of the already-mentioned question which is the simple probability analogue of (Q2). To begin our answer with "(Because) it is a scientific law that ... ", or with "(Because) to date it has always turned out that ... ", or, in general, with "(Because) contingently ... " would be to renounce all hope of finding an answer to (Q2) analogous to (A1). For it would be to admit at the outset that, whereas it is of course necessarily true that one should believe what is certain in relation to one's total evidence, it is simply *not* necessarily true that one should believe what is probable but not certain in relation to that evidence. If this were to be admitted, then the only proper response to (Q2) would be, not to try to answer it in analogy with (A1), but to reject the question as containing a most serious *suggestio falsi*, and one to which (Q1) seems to contain no counterpart.

A second feature of (A1) which must be regarded as fixed also for the answer to (Q2) is its continuing with "(Because) (necessarily) *if E is true then* ... " Or, at any rate, I must regard this feature as fixed. This is simply because, although it is hard to complete an answer to (Q2) which contains this proviso, without this proviso I simply have no idea at all of the way an answer to (Q2) might go after "because necessarily ... ".

Now (A1) ran thus: "Because, necessarily, if E is true, then *every H* (which stands in the specified relation to E) is *true*." Well, with the first seven words fixed, as being needed also in our answer to (Q2) and with the specified relation varied from certainty to probability, what possibilities remain for making other variations of (A1) suitable for an answer to (Q2)? There seem to be only two. One could leave the (final) "true" intact but vary the "every" along the proportion-dimension. That is what (A2″) did. Or one could leave the "every" intact, and try to vary the "true" – and how else but along the simple-probability dimension? That is what (A2′) did. Either course, as we have seen, is radically objectionable. But, as far as I can see, there are no other variations which could be made at either of these two places, and no other place at which any variation could be made. That is why I said that in the answers to (Q2) already found wanting there is at least an approach to exhaustiveness.

Our negative results, therefore (taken along with the assumption made in favour of (A1) at the outset), suggest the conclusion that whereas there is an adequate answer to (Q1), viz., (A1), there is none at all to (Q2).

This conclusion is alarming. For one thing, it would seem to involve us in Hume's notorious scepticism concerning induction. All inductive arguments are fallible; that is, it is possible for their conclusions to be false, even though their premisses be true. Consequently, the conclusion of an inductive argument is at best probable, never certain, in relation to the premisses. But we now seem unable to answer someone who asks us why one should believe conclusions which are probable but not certain and, hence, unable to answer someone who asks us why one should believe the conclusion of any inductive argument. And is not this to be unable to answer Hume's inductive scepticism?

The conclusion that (Q1) is adequately answered by (A1) while (Q2) is not answerable in any analogous way, seems especially alarming for those philosophers who adhere to the Keynes—Carnap theory of probability, the "logical-probability" theorists; or perhaps only for them. For (A1) answers (Q1) in terms of the truth-frequency among propositions H which are certain in relation to E; whereas, no answer to the (A2″) type, that is, no answer in terms of the truth-frequency among propositions H which are probable in relation to E, appears to be available for (Q2). This lack of symmetry would seem to show that the logical theory of probability provides "no basis for expecting the probable rather than the improbable . . . and [that it] lacks predictive content and thus fails to qualify as a 'guide of life'."[2]

This suspicion is confirmed by certain remarks of Keynes himself, at the scattered points in *A Treatise on Probability* where he touches on what is, in effect, my (Q2). Keynes's answer to (Q2) appears to be the "repudiationist" answer (A2) above. A conclusion which is probable but not certain in relation to one's total evidence has – such seems to be Keynes's view – *"nothing to recommend it but its probability"*.[3] This is cold comfort indeed, and the contrast between this answer to (Q2) and the answer (A1) to (Q1) is painfully marked. Probability, then, is to be, like virtue according to the Stoic doctrine, its own reward. Stoic doctrine indeed, of which the plain English is that there is *no* reward for believing what is probable! For believing what is certain in relation to one's total evidence, by contrast, (A1) holds out the reward, concrete though conditional, of truth.

II

What I have done in Section I above is to give my own version of an argument which Professor Wesley Salmon, if I have understood him rightly, has several times advanced.[4] I should stress that this version is my own, and therefore may contain some inadvertent misrepresentation.

In the present section I will try to show that in two important respects Salmon has misconceived the conclusion to which his argument points. In the next section, I return to examine the starting point of the argument, the questions (Q2) and (Q1).

Salmon appears to think, as I have indicated, that the difficulty of answering (Q2) in analogy with (A1) is a difficulty especially for, or perhaps only for, the theorists of logical probability. But it should be obvious that this is not so. There is absolutely nothing in the argument of the preceding section which depends in any way at all on any thesis which is peculiar to the theorists of logical probability. (In particular, their thesis that assessments of the probability of one proposition in relation to another are non-empirical propositions has played no part whatever in the argument, either overtly or covertly.) This is something which the reader can easily verify for himself by referring back to Section I.

In order to be exposed to the difficulty of answering (Q2) on lines analogous to (A1), one does not need to be a logical-probabilist, or even a philosopher. All one does need to be, in fact, is someone who thinks that one *should* believe what is probable though not certain in relation to all the evidence one has. And who thinks that? Why, nearly everyone. The difficulty which Salmon has brought to light then, if it is a genuine one, is equally a difficulty for everyone, and certainly not only or especially for one tiny group of philosophers.

A second and much more important misconception which Salmon entertains, if I have understood him rightly, is that the difficulty of finding an answer to (Q2) which is analogous to (A1) "is Hume's problem of induction once again".[5] It is nothing of the kind.

An inductive argument, according to the main stream of philosophical usage, is simply an argument from observed to unobserved instances of some empirical predicates. Now, in that conception, it is important to notice, there is nothing which precludes the conclusion of an inductive argument from being certain in relation to the premisses; and, indeed, nothing whatever about the degree of probability which the conclusion of an inductive argument can or cannot have in relation to the premisses. Equally, in the conception of an argument whose conclusion is (at best probable but) not certain in relation to the premisses, there is nothing to tell us that there is any *inductive* argument which falls under that conception. If, then, someone who asks (Q2) wishes to bring inductive arguments within the scope of his question, it will be necessary for him to do what was done in the third-last paragraph of Section I, viz., to *state* that the conclusion of an inductive argument is at best probable, never certain, in relation to the premisses.

Now, this proposition, despite what Carnap, Edwards, and many others have implied to the contrary, is not a triviality like "Bachelors are unmarried". It is a logico-philosophical thesis, and an extremely important one: the thesis of "inductive fallibilism" as I call it. I think indeed that this thesis is true, as most philosophers now think; though it has not always been thought so.[6] But even if it were not true, we would have no less need to answer (Q2), and no less difficulty in doing so, than we have as things are. Let us suppose that arguments from observed to unobserved instances of empirical predicates are all such that their conclusions are certain in relation to their premisses. Clearly, that is not going to make the possible answers to (Q2) any more numerous, or any less defective, than we have found them to be! That supposition would deprive us of what is perhaps our most important class of *examples* of arguments, whose conclusions are not certain in relation to their premisses; but that is all it would do. And that would be no fatal deprivation. When we wished to illustrate (Q2), we could simply draw all our examples from some non-inductive class of fallible arguments. This is, in fact, what I did do above. The lottery arguments, and the arguments about Tex, are arguments none of whose conclusions are certain in relation to their premisses; but none of them is inductive. They belong, on the contrary, to what was formerly, and aptly, called "direct" probability, as expressly distinguished from "inverse" or "inductive" probability. (In particular they were examples of what may aptly be called "Bernoullian" arguments.)

The question (Q2), which we have found difficulty in answering, then, is far from being about inductive arguments only, or about them especially, because it is not necessarily about them at all. It is safe to assume that, on the other hand, Hume's scepticism about inductive arguments *is* necessarily about inductive arguments. Consequently, the difficulty of answering (Q2), whatever it is, is at any rate *not* the difficulty of answering Hume's inductive scepticism.

This is still putting the matter much too mildly.

Hume's inductive scepticism, it is reasonable to assume, is some proposition, some thesis or other. (Q2), on the other hand, is not a proposition but a question.

More important still: Hume's inductive scepticism, as I have tried to show in detail elsewhere,[7] is actually a version of the thesis that the conclusions of inductive arguments, in relation to their premisses, *are not probable.*[8] Now, the mental state which that thesis expresses is very different from that which is expressed by the question (Q2), and even incompatible with it. For the mental state behind (Q2) is that of someone who concedes that sometimes a proposition *H* is probable though not certain in relation to his total evidence *E*, and who could with perfect consistency admit that some of these cases happen to be ones in which the argument from *E* to *H* is inductive; but who then goes on to ask, concerning all such cases (inductive or not) indifferently, why he should believe any such *H* since any such *H* may be false. It is surely obvious that this person is maintaining a different position from that of someone who just denies that the conclusions of inductive arguments ever are probable in relation to their premisses.

But I do not need to rely only on indirect arguments, for the matter before us has a straightforward textual side. If we are to be persuaded rationally that (Q2) is an expression of Hume's inductive scepticism, it can only be by references to Hume's text which are sufficient to sustain that attribution. Although Salmon, if I understand him rightly, does attribute (Q2) to Hume, he gives no such references; indeed he gives no references which are even intended specifically to support that attribution. A sufficient reason is that no such references could be given. Not only is (Q2) not an expression of

Hume's inductive scepticism, it is not an expression of anything whatever to be found in Hume's writings. The attribution of (Q2) to him is simply a glaring instance of a bad habit of twentieth-century philosophers, of which I have elsewhere collected some other examples, of fathering on Hume things which are no more than the sprouts of their own brains.[9]

How this particular cuckoo's egg can ever have looked at home in the Hume nest is hard for anyone with any feeling for the history of thought to imagine. Hume's inductive scepticism is clear, and straightforward, and a thesis; and in these and many other respects it manifestly "belongs" where it actually occurs, viz., at the centre of the Scottish Enlightenment of the eighteenth century. (Q2), on the other hand, has a contrived and morbid air which stamps it as a question which equally belongs where *it* actually occurs, viz., in the mid-twentieth century and nowhere else.

It will be worthwhile briefly to enlarge on the state of mind which (Q2) expresses, and to ask after its probable cause. And here, despite what I have said above, I think it is true, *causally* speaking, that Hume's philosophy of induction will be found at the bottom of the matter.

The *ground* on which (Q2) is asked − and I have incorporated this in the question all along − is that any *H* which is probable but not certain in relation to one's evidence *E may be false*. Yet that, of course, was given in the question, since to say that *H* may be false is just another way of saying that *H* is not certain in relation to *E*. It was equally given in the question, on the other hand, that the *H*s in question *are* probable in relation to *E*. But this part of the data is simply lost on a person who asks (Q2) in earnest. He is someone who can think only of the other part, the possibility that *H* is false. "Still, I *may* win all the lotteries", "Still, the sun *may* not rise" − these are the thoughts which occupy his mind to the exclusion of all else. This questioner is someone anaesthetised to probability, hyperaesthetised to possibility: he is "probability-numb".

This sounds like a rather strange mental state, and certainly a distilled expression of this state, such as (Q2), is not to be met with every day. Yet the state itself seems to me to be not really an unfamiliar one nowadays. Salmon's question, though new, has not lacked other philosophers to second it. But much more than that, large areas of probability-numbness seem to me to exist at present in our intellectual culture generally. The philosophy of science is one such area. The feature of scientific arguments on which nowadays we are most anxious to insist, and for many of us the sole feature on which we are at all eager to insist, is their fallibility: the permanent possibility that the conclusions may be false, even though all the empirical evidence for them be true. We know that by concentrating so exclusively as we do on this feature, we open the door to an extreme irrationalism about science (Feyerabend's for example), which we do not particularly welcome. But still, on the whole, we stay numb to any aspect of scientific inference except its fallibility. A more complete contrast with the philosophers of science of 1876 or of 1776, could scarcely be imagined; for where they characteristically made too favourable an assessment of the probability of scientific arguments, and prized scientific conclusions chiefly for their supposed certainty, we characteristically avoid making *any* assessment of the probability of those arguments, and prize scientific conclusions for – what reason is not quite clear. And what brought this change about was not that we quietly outgrew "the search for certainty"; though we think we did, which is why that phrase is now found so incomparable a soporific. No, we are separated in our philosophy of science from the nineteenth and eighteenth centuries rather by this: that we suffered sudden certainty-*deprivation*, and are to this day still in the resulting state of shock, a state in which mere probabilities leave one cold.

And who brought on the shock? Who taught twentieth-century philosophers of science, what their predecessors saw only fitfully or dimly or not at all, that even the best inferences from empirical evidence to scientific conclusions are incurably fallible? Taught it to us with such overwhelming force, I may add, that by 1950 it was often made, what it never was before, part of the *meaning* of the phrase "inductive inference", that it be possible for the conclusion to be false and premisses true; and even made, by Carnap [10] and most of his followers, the *whole* meaning of "inductive inference"?

Why, David Hume of course; with the assistance of Einstein, who (though himself an avowed Humean fallibilist) enabled philosophy to teach by example, and that example, the most resounding of all. And what philosophy was irresistibly taught by the ending of the *pax Newtoniana*? Not Hume's inductive *scepticism*, although naturally in the ensuing period of turbulence a few *esprits forts* have been found willing to embrace even that. It was Hume's inductive *fallibilism*.

This seems to me to be the historical key to (Q2). It is the realisation that *whatever* the weight of empirical evidence in its favour, any scientific theory *may* be false – it is that *possibility*, brooded on to the exclusion of every other feature of scientific inference – which has brought (Q2) before us. This strange and novel question is a delayed effect of the intellectual earthquake set off early in the twentieth century by Hume and Einstein, a faint tremor reaching us from that great but distant subsidence of scientific

confidence, and now amplified by morbidly nervous philosophic ears. Such
at any rate appears to me to have been the aetiology of (Q2).

Inductive fallibilism is only one of Hume's profound and influential *non-
deducibility theses*. Two others are: the non-deducibility of factual propositions from necessary truths ("There can be no demonstrative arguments for a
matter of fact and existence"), and the non-deducibility of "ought" from "is".
But non-deducibility theses are especially abundant in his philosophy, and
are, in fact, chiefly what give it its negative or destructive character. That his
philosophy *is* of a predominantly destructive character, I take to be as obvious
to Hume's readers now, even after the efforts of Professor Kemp Smith to
portray it in an opposite light,[11] as it always was before. Indeed, when I consider the immense destructive effects which have actually been wrought, by
the three theses just mentioned, on scientific confidence, religious confidence,
and moral confidence, respectively, I wonder whether the philosophy of Hume
is not at the bottom, not just of probability-numbness in our present philosophy of science, but of the far wider phenomenon of "modern nervousness",
of which Freud advanced a different and singularly improbable explanation.[12]

To recapitulate. The difficulty of answering (Q2) in analogy with (A1)
is not a difficulty only or especially for the logical theory of probability.
Nor is this difficulty "Hume's problem of induction once again". No doubt
someone who asks (Q2) in earnest can rightly be called a sceptic of *some*
kind. But his kind of scepticism is not Hume's, if only because it has no
necessary connection at all with inductive arguments. His kind of scepticism
need not even extend to inductive arguments, though it may do so. It will
do so, if he is an inductive fallibilist (and is consistent). But even if it does,
his scepticism about inductive arguments will still be quite different from
Hume's scepticism about them; for Hume's consists in denying that their
conclusions ever are probable in relation to their premisses, whereas the (Q2)
kind of scepticism could admit that some such conclusions are probable in
relation to their premisses, but goes on to ask, in effect, "So what?"

But at best all this shows only that Salmon has misinterpreted the difficulty
of answering (Q2) in analogy with (A1). We have yet to dispose of that difficulty itself.

III

The answer (A1) to (Q1) states a necessary connection between certainty
in relation to evidence, and truth and, hence, it ensures a certain truth-frequency among some of our conclusions. Clearly, what Salmon and indeed

all of us would *like* most, by way of an answer to (Q2), is something similar for the case of probability: that is, an answer which states a necessary connection between probability in relation to evidence, and truth, and which therefore ensures a certain truth-frequency among some of our other conclusions. An answer of this *type* has, of course, already been canvassed above, viz., (A2″). But (A2″) stated only a very simple form which the desired connection might take, and anyway (A2″) was false. What is wanted is something which, like (A1), states a necessary *truth*; only, one connecting probability with truth.

This, which I shall call "Salmon's desideratum", is so natural a one that it will be worthwhile to prove that it cannot be satisfied. It cannot, because its satisfaction would be inconsistent with the satisfaction of another, and genuine, desideratum for an adequate answer to (Q2).

Suppose (Q2) were given an answer which, although true, was such that it never, when added to one's total evidence E, raised to certainty the probability of any H which was not certain in relation to E alone. Such an answer would not be adequate. For such an answer, if it were added to the total evidence E of the questioner in particular, would *ex hypothesi* leave every proposition H which was not certain in relation to his E before not certain still. Now our questioner's ground for asking why he should believe an H which is probable but not certain in relation to his E was that any such H may be false. But any H of his which before might have been false still might be false after he has added to his E an answer to (Q2) such as we are at present supposing. Given such an answer, then, our questioner, unless he were just inconsistent, would have to renew his original question (Q2). But if an answer to one's question, when added to one's total evidence, leaves one with exactly the same ground for asking the question as one had before, it is inadequate. On the other hand, an adequate answer to (Q2) must be such that there is at least some H and some E such that while H is probable but not certain in relation to E, H *is* certain in relation to E plus that answer.

But this desideratum and Salmon's cannot be met together. Any E in relation to which some H has probability < 1 is consistent, and any H which in relation to some consistent E has probability < 1 and > 0 is contingent. Hence H must be contingent and E consistent in order to satisfy the first part of the desideratum just stated. But then the second part of that desideratum cannot be satisfied along with Salmon's. For it is impossible to raise to certainty the probability of a contingent proposition H by adding, to consistent evidence E, a necessary truth connecting probability with truth, or by adding any necessary truth whatever. (To suppose otherwise would be to suppose that with N necessarily true, H contingent, and E consistent, it is possible that (i) $P(H/E) < 1 > 0$ and (ii) $P(H/E \cdot N) = 1$. But if (i) is true, then the conditional $E \supset H$ is neither necessarily true nor necessarily false, hence must be contingent; while if (ii) is true, that contingent proposition is certain in relation to N alone, which is impossible.)

It will be worthwhile to illustrate what this argument shows, and best to do so by reference to the special case of inductive arguments. The argument shows this: that if we *could* get an answer to (Q2) of the type which Salmon and the rest of us would most like, viz., a truth stating a necessary connection between probability and truth, this answer, even when we added it to our total evidence E, would leave every single inductive conclusion H which was not certain before not certain still. All our generalisations and predictions from experience, in other words, even the most probable ones, would *still have nothing to recommend them except their probability*!

The same argument also proves a more general result: that two of the desiderata for an adequate answer to (Q2) are inconsistent. These are: the desideratum that an answer to (Q2) be a necessary truth; and the desideratum that it be sufficient, when added to some E, to make certain some H which in relation to E alone was not so.

This consideration ought at least to incline us to the view that (Q2) is a question to be repudiated rather than answered. I can render this view more acceptable still by pointing out a certain defect in *all* the answers we have considered to (Q2), and another one in that question itself.

In Section I, I said that an adequate answer to (Q2) must, after "Because necessarily", continue " ... if E is true ... "; though I admitted that I had said this only because I could conceive of no other way in which an adequate answer could continue.

I now point out, however, that even if one could complete in an otherwise adequate way an answer beginning so, the result would be at most *half* of an answer to (Q2). For the obligation to believe what is probable in relation to one's total evidence E is not confined to the cases in which E is true. It subsists also and equally where E is false. A second arm of an answer is required, therefore, one which begins " ... ; and if E is false ... " But now, I think it will be agreed, no one has any idea at all how to complete the part of an answer to (Q2) which begins in that way. Least of all, does anyone know how to complete it in terms of the *truth-frequency* among Hs probable in relation to E. (For nothing can be said *a priori* and in general about the truth-frequency among propositions probable though not certain in relation to false E.) And even if this could be done, the result would still be not quite the kind of answer which is required. For the obligation to believe what is probable though not certain in relation to one's total evidence E has really nothing at all to do with the truth *or* the falsity of E. An answer to (Q2) accordingly, should take account, not of both, but of neither of those cases. Yet no one, I think it will be agreed, has any idea how an answer to (Q2) which takes no account of the truth-value of E might run. Least of all does anyone know how an answer, which is in terms of the truth-frequency among Hs, might run once the truth-value of E is disregarded. (For nothing can be said *a priori* and in general about the truth-frequency among propositions probable though not certain in relation to an E which is of unspecified truth-value.)

If all this is so, then we really have no idea how to answer (Q2) at all.

The question itself, moreover, contains a *suggestio falsi* of the most serious kind. It suggests that always, when one believes what is probable though not certain in relation to one's total evidence, one believes as one should, and it asks why this is so. But it is *not* so. What is true is that if H has probability $= x/y$ in relation to one's total evidence E, where $x/y > 0 < 1$, one should have in H a degree of belief which is a fraction $= x/y$ of one's degree of belief in E. But if, for example, H has probability $= 0.9$ in relation to my E, and I have in H a degree of belief which is a fraction < 1, or $= 0.8$, say, of my degree of belief in E, then while indeed I believe *what* I should, viz., H, it is by no means true that I believe *as* I should. Nor is this false suggestion only of peripheral importance in the present context. On the contrary. The very thing which assessments of probability do, and which no other propositions do, is to characterise certain *degrees* of belief as rational, and others as not , as distinct from characterising just some *beliefs* as rational, or not.

It would be necessary, in order to free (Q2) from this *suggestio falsi*, to reformulate the question as follows:

(Q2′) Why should one have, in a proposition H which has probability $= x/y$ in relation to one's total evidence E (where $x/y > 0 < 1$, but may be close to 1), a degree of belief which is a fraction $= x/y$ of one's degree of belief in E?

And, now, to *this* question what answer could possibly be given, except some repudiationist one? What could one say to the questioner, except, in the spirit of Keynes and of (A2), "You *told* us why one should! Because H has probability $= x/y$ in relation to E."

But neither is (A1) an adequate, even if it is an appropriate, answer to (Q1). It is at best *half* of such an answer. For the obligation to believe what is certain in relation to one's total evidence E is not confined to the cases in which E is true. It subsists also and equally where E is false. A second arm of an answer is required, therefore, one which begins " ... ; and if E is false ... ". But, now, I think it will be agreed, no one has any idea at all how to complete the part of an answer to (Q1) which begins in that way. Least of all does anyone know how to complete it in terms of the truth-frequency among Hs certain in relation to E. (For nothing can be said *a priori* and in general about the truth-frequency among propositions certain in relation to false E, except that it is < 1.) And even if this could be done, the result would still be not quite the kind of answer which is required. For the obligation to believe what is certain in relation to one's total evidence E has really nothing at all to do with the truth *or* the falsity of E. An answer to (Q1), accordingly, should take account not of both but of neither of those cases. Yet no one, I think it will be agreed, has any idea how an answer to (Q1), which takes no account of the truth-value of E, might run. Least of all, does anyone know how an answer, which is in terms of the truth-frequency among Hs, might run once the truth-value of E is disregarded. (For nothing can be said *a priori* and in general about the truth-frequency among propositions certain in relation to an E which is of unspecified truth value.)

If all this is so, then we really have no idea at all how to answer (Q1), either.

But that very question, like (Q2), contains in addition a *suggestio falsi* of the most serious kind. It suggests that always when one believes what is certain in relation to one's total evidence, one believes as one should, and asks why this is so. But it is not so. What is true is that if H has probability $= x/y = 1$, i.e., is certain, in relation to one's total evidence E, one should have in H a degree of belief which is a fraction $= x/y = 1$ of one's degree of belief in E. But if, for example, H has a probability $= 1$ in relation to my E, and I have in H a degree of belief which is a fraction < 1, or $= 0.8$, say, of my degree of belief in E; then while indeed I believe *what* I should, viz., H, it is by no means true that I believe *as* I should. Nor is this false suggestion only of peripheral importance in the present context. On the contrary. The very thing which assessments of probability do (assessments of probability $= 1$ or certainty among them), and which no other propositions do, is to characterise certain degrees of belief as rational, and others as not; as distinct from characterising just some *beliefs* as rational, or not.

It would be necessary, in order to free (Q1) from this *suggestio falsi*, to reformulate the question as follows:

(Q1′)   Why should one have, in a proposition H which has a probability $= x/y = 1$, i.e., is certain, in relation to one's total evidence E, a degree of belief which is a fraction $= x/y = 1$ of one's degree of belief in E?

To *this* question, it will be evident, the original answer (A1) is not only not adequate, but not even appropriate. And what answer could possibly be given to (Q1′), any more than to (Q2′), except a repudiationist one? "You told us why one should. Because H is certain in relation to E."

To summarise, an adequate answer to (Q2) could not be a necessary truth connecting probability with truth.

An adequate answer to (Q2) would have to satisfy inconsistent conditions.

The answer (A1) to (Q1), and all of the possible answers to (Q2) analogous with (A1), are inadequate, in not reflecting the fact that the obligations to believe what is certain and to believe what is probable in relation to all one's evidence are both independent of the truth-value of that evidence; and it does not seem possible to remedy this defect in either case.

Both (Q1) and (Q2) contain serious *suggestiones falsi*, the avoidance of which would turn these questions into (Q1′) and (Q2′), *neither* of which is a question admitting of a substantial answer any more than does "Who wrote the novels Scott wrote?" or "What number is three?" Thus the contrast between the apparent answerability of (Q1) and the apparent unanswerability of (Q2) disappears along with those questions themselves.

IV

But someone who asked my original question (Q2) might well mean by it something quite different from the question which in Section III we have seen reason to repudiate. I can think of three such things that (Q2) might mean. In this final section I discuss these.

"Why should one believe ... " sometimes means the same as "What evidence is there for believing ... ". Consequently (Q2) might mean: "What evidence is there for believing any of those propositions H which are probable though not certain in relation to one's total evidence E?"

This is clearly a very foolish question. I do not accuse any actual person of having intended to ask it. Yet I am not sure that some of the modishly "Sceptical" questions I have seen in print – though usually arbitrarily confined to the special case of induction, and almost always groundlessly fathered on Hume – did not mean the above (perhaps among other meanings). Anyway

(Q2), if it ever does mean this, deserves to be answered even more unsympathetically than on any other interpretation of it. For the answer to the above question is celarly just: "E!" And to an *inductive* scepticism, in particular, which was of this degenerate variety, the correct answer would consist just in a report of past experience.

A person might ask (Q2) with, as it were, invisible scare-quotes around the word "probable": he might be wondering whether what *passes with us* for probable really is so. Thus a second possible meaning for (Q2) is this: "Prove that the propositions $P(H/E) = x/y$ $(0 < x/y < 1)$, which are naturally or usually *believed*, are true."

It must be admitted that there are cases, that is, particular values of H, E, and x/y for which this demand cannot be satisfied at any time. A particular "natural" assessment of probability $P(H/E) = x/y$ may be, although natural, false; and on that account permanently incapable of proof. Another particular assessment $P(H/E) = x/y$ may be true as well as natural, and yet be temporarily or permanently incapable of being proved, either through our ignorance or stupidity, or because this particular assessment of probability is simply one of those which cannot be derived from anything else whatever.

Yet it is certain, too, that there are cases even at the present time, in which the above demand can be satisfied. How much can, in fact, be done in the way of such proofs is much less well known among philosophers than it deserves to be.

Consider for example the following assessment of the probability of a certain Bernoullian inference:

(A) $P(a$ is red/Just two out of a, b, and c are red$) = \frac{2}{3}$.

It is an eminently natural assessment. But it might be thought, and has in fact been thought by good philosophers, to be one of those natural assessments of probability which are destined to remain permanently underived from anything else. Yet one of the things which Carnap made known in 1950 was that that is not so; that on the contrary (A) can easily be derived from extraordinarily weak premises of just the following two kinds.

(B) $P(a$ is red and b is red and c is not red/$T) < 1$

and

(C) $P(a$ is red/$T) = P(b$ is red/$T)$,

where T is some tautology.[13] So if our questioner were to demand a proof, for example, that the natural assessment of probability (A) is also a true assessment, then his demand could be satisfied at present, and with the greatest *éclat*. (For even the most inveterate opponent of what Carnap calls "inductive|logic" will not contest the truth of premisses of either the (B) or the (C) kind. The symmetry of individual constants [that is, their uniform exchangeability *salva probabilitate*], which is sufficient for (C), is so indispensably necessary, even for deductive logic, that no one could afford to deny (C). And (B) is a non-deducibility thesis of the most obvious possible kind and, hence, a proposition which the more hardened a "deductivist" one is, the more one will heartily affirm.)

A demand for proofs of the truth of natural assessments of probability, then, is in some cases able at some times to be met, though in other cases or at other times it will not be able to be. Whether or not it can be met in a given case will depend partly on the state of our knowledge, and partly on which particular natural assessment a proof is demanded for.

This shows, however, that the above demand is a perfectly fair and limited demand, which does not express any kind of scepticism whatever about assessments of probability. It is as innocuous as a demand, say, that natural *arithmetical* beliefs be proved to be true: a demand which no doubt can be met in some cases at some times though not in all, and which expresses no kind of scepticism about arithmetical propositions.

The question (Q2), therefore, if and when it means just this demand, is likewise perfectly innocuous. It expresses no kind of scepticism about assessments of probability, and, hence, need occupy us here no further.

What *would* express a kind of scepticism would be a demand that *every* assessment made of the probability of one proposition in relation to another be proved. It is certainly impossible to satisfy *this* demand. For an assessment of the probability of one proposition in relation to another can be validly derived only from premisses which themselves include other assessments of the probability of propositions in relation to others. It must not be supposed that there is anything sinister in this. The same is true not only of the kind of probability which we are here concerned with, but of any kind of probability at all. For example, assessments of "factual" probability or of long-run relative-frequency (Carnap's "probability$_2$") can be validly derived only from premisses which themselves include other assessments of factual probability or of long-run relative-frequency. Still, it *is* true of the probability of one proposition in relation to another, as well as of all other kinds of probability. And a demand that every assessment made of this kind of probability be proved would therefore be, in effect, a demand that no assessments of this kind of probability ever be made at all.

Just that, though, is clearly a third possibility as to what (Q2) might mean. Earlier I compared (Q2) with "Why should one respect what is sacred?" Now that question *might* mean: "Why should I not just stop ascribing sacredness to anything? What (intellectual) trouble would I get into if I simply let the concept of the sacred drop out of my life?" Well, similarly (Q2) might mean "Why should I not just stop assessing the probability of propositions in relation to others? What (intellectual) trouble would I get into if I simply let this concept of probability drop out of my life?"

This question about the sacred seems to me to be not only perfectly proper; I do not believe its challenge can be met. *No* intellectual trouble results if one drops the concept of the sacred, and one should drop it. The above question about probability, too, is perfectly proper; but its challenge can be met. The intellectual trouble which a poser of (Q2) gets into, if (Q2) means this challenge, is no less than inconsistency.

My poser of (Q2), it will be recalled, asked that question on a specific ground: the ground that any H probable but not certain in relation to its total evidence E may be false. But now, what does his "$H \ldots$ *may be* false" mean? Not, for example, what in another context it might mean: that H is not a necessary truth. No, that H may be false clearly means here that the falsity of H is possible in *relation to E*. But to say that is no different from saying that the probability of H in relation to E is < 1. And to say that is to make a certain assessment of the probability of one proposition in relation to another. If, therefore, someone who asked my (Q2) meant by it a proposal to refrain from making assessments of the probability of propositions in relation to others, then his proposal would be inconsistent with the ground on which he makes it. For that ground itself consists of assessments of the probability of propositions in relation to others; assessments, namely, of the $P(H/E) < 1$ kind.

"But there is a great deal of difference", Professor Popper for one would object here, "between assessments of probability of the $P(H/E) < 1$ kind, and those of the $P(H/E) = x/y$ kind, where $0 < x/y < 1$. The former are completely innocent, and can even be discovered to be true by means of empirical counter-examples. It is the latter kind of assessment alone which one should refrain from making. For that kind is admitted by Carnap to rest in the end on intuition; and one should not rely on intuition. If your charge of inconsistency were correct, one could not so much as point out a fallacy – and that is all that making the $P(H/E) < 1$ kind of assessment comes to – without getting oneself implicated in the number-magic of so-called inductive logic. But this is absurd."[14]

Unfortunately for this reply, assessments of the kind Popper objects to, $P(H/E) = x/y$ where $0 < x/y < 1$, are in some cases *derivable from* assessments of kinds he does not object to. The derivability of (A) above from premises just of the (B) and (C) kinds is an example. But more importantly still, with trifling exceptions, reliance on intuition is required in learning the truth *even of assessments of the $P(H/E) < 1$ kind.*

It will be possible to learn the truth of $P(H/E) < 1$ purely empirically: if E and H are constants (i.e., if $P(H/E) < 1$ assesses the probability of a concrete argument, not of every member of a certain *class* of arguments); if E is true; if E is observational; if not-H is true; if not-H is observational; if E-and-not-H is observational. For then it will be possible to discover that E-and-not-H is *possible*, by discovering empirically that it is true. But if any one of the above exceedingly restrictive conditions fails, then it will not be possible to learn the truth of $P(H/E) < 1$ empirically.

For example, it is impossible to learn, from experience alone, even so meagre a fragment of elementary logical truth as:

(D) $P(x$ is a present member of the Politburo/All present members of the Politburo are men and x is a man$) < 1$.

Observational counter-examples do, of course, exist to the class of arguments every member of which is correctly assessed by (D) as being fallible. Indeed, they abound, since each man not a present member of the Politburo is one such counter-example. But the truth of (D) could not be learnt just by observing one or any number of those counter-examples. Such observation would not teach us the truth of (D) unless it were supplemented by knowledge of the *symmetry of individual constants*. We all do have such knowledge, of course. But no one will maintain that this knowledge is empirical.[15]

If even so small and uninteresting a non-deducibility thesis as (D) cannot be learnt without at least some reliance on intuition, still less can large and interesting ones be learnt without it. The non-deducibility theses of Hume, for example, mentioned earlier, and the thesis of inductive fallibilism among them, certainly cannot be learnt without it. Those assessments, and almost every other assessment of the $P(H/E) < 1$ kind, are quite as empirically inaccessible as assessments of the $P(H/E) = x/y$ kind where $0 < x/y < 1$.

The inconsistency which I pointed out above, then, cannot be escaped by pleading that assessments of probability of the $P(H/E) < 1$ kind are "only little ones". They are only little ones, in the sense that $P(H/E) < 1$ is a far weaker proposition than, for example, $P(H/E) = \frac{2}{3}$. But, still, assessments of probability are what they are, and with the trifling exceptions noticed above, the truth of $P(H/E) < 1$ can be learnt only in the same non-empirical way as the truth of $P(H/E) = \frac{2}{3}$.

(Q2), if it means the first of the things considered in this section, is foolish; if it means the second, it is innocuous; if it means the third, it is inconsistent. And these are all the things I have been able to think of that (Q2) might mean, if it were not the question which we saw reason in Section III to repudiate.

The inconsistent position which was discussed a moment ago as a possibility is at least perilously close to the *actual* position of most contemporary philosophers of science, if what was said of the latter in Section II is true. I said there that these philosophers systematically refrain from assessing (though especially from assessing favourably) the probability of scientific conclusions in relation to their evidence, and that the historical *cause* of their so refraining is chiefly the influence of Hume's inductive fallibilism. Now, if that is true, then all that would be needed to make their position the inconsistent one which was discussed above is that inductive fallibilism (as well as being the cause of it) should be taken as a *ground* for not assessing the probability of scientific conclusions in relation to their evidence. For that would be, again, to take certain assessments of probability, of the $P(H/E) < 1$ kind, as a ground for not assessing probabilities.

Well, *is* the truth of inductive fallibilism ever taken as a ground for not assessing probabilities in contemporary philosophy of science? It seems to me to be constantly so taken. I am not aware, for example, of any *ground* of the Popperians' hostility to the Carnapians' proposal to assess the probability of inductive arguments, except in the end just this, that inductive arguments *are* fallible! But I will not lengthen an already long essay by trying now to document this accusation. It will be safe, though, to speak hypothetically: *if* the mere possibility of a false conclusion with true premisses is widely taken as a ground for not assessing the probability of scientific arguments, then our current philosophy of science contains a massive inconsistency. We would be no better than, in fact we would be exactly like, another group of contemporary irrationalists, the antinomian pantheists, to whom, grotesquely, nothing is sacred *on the ground that everything is*. And just as those persons, while thinking they have dispensed with the concept of the sacred, are secretly those who are most intoxicated with it, so we, while thinking we never assess the probability of scientific arguments, would secretly be, in fact, the persons who most constantly and even obsessively do so.

University of Sydney

NOTES

[1] Reprinted by permission of the publisher from *Hume: A Re-Evaluation*, edited by D. Livingston & J. King (New York: Fordham University Press, 1976), Copyright © 1976 by Fordham University Press, pp. 50–68.

[2] Salmon ([1967a], p. 79)

[3] Keynes ([1921], p. 322); emphasis added.

[4] See Salmon ([1967a], pp. 48–49, 74–79); ([1965a], pp. 265–270, 277–80); [1968b].

[5] Salmon ([1967a], p. 79); see also p. 52. The same is said in Salmon ([1968b], p. 33).

[6] I have tried to trace some of the historical fortunes of this thesis in my [1973], Chap. 8.

[7] Ibid., Chaps. 2–4.

[8] Hume's inductive scepticism, if I have understood it rightly, is an assessment of probability belonging to the comparative kind which Keynes called "judgments of irrelevance". In particular, it is the thesis that for all tautological E, and for all E' and H such that the argument from E' to H is inductive, $P(H/EE') = P(H/E)$. That is, it asserts that the conclusions of inductive arguments are not even more probable, in relation to their premises, than they are in relation to tautological evidence alone. This is clearly a very unfavourable assessment of the probability in question, and inconsistent with describing the conclusions of inductive arguments as being, in relation to their premises, "probable" in any sense of that vague but still unmistakably *favourably*-evaluative word. (Much as, if a man is not even taller than the average man, he is *a fortiori* not "tall".)

[9] For other examples, see Stove ([1973], pp. 125–132). To these might be added the following, from Salmon ([1967a], p. 52) (emphasis added): "*As Hume has shown*, we have no reason to suppose that probable conclusions will often be true and improbable ones will seldom be true."

That is sheer invention masquerading as history. And unfortunately this is a variety of mischief which is far easier to do than to undo. What indeed can one do against such a thing, except give it a flat denial? I say, then, that Hume did not show what Salmon here says he showed; that he never attempted to show it; that it never occurred to him to attempt to show it.

Between Salmon's assertion and my counter assertion, the matter now rests entirely with the texts. In order to prove the truth of his assertion and the falsity of mine, all Salmon needs to do, though also what he must do, is to put before us those passages in which Hume showed what he is here claimed to have shown.

For Salmon, or anyone who wishes on his behalf to accept this challenge, perhaps the least unpromising place to start, would be Book I, Part III, Section XI, Paragraph VIII of the *Treatise* – though Salmon has never referred to that passage, or, indeed, to the *Treatise* at all. But they should also be warned that up to the present no writer on Hume has been able to get out of that paragraph any clear philosophical matter at all, let alone anything important. The same is true, indeed, of this baffling Section XI as a whole (apart from the two introductory paragraphs and interspersed repetitions of material from earlier sections of Book I, Part III).

[10] See Carnap ([1962], p. 580) (Glossary, under "inductive inference") and *passim*. This twentieth-century change in the meaning of "inductive" has had several effects, all bad. One is the trivialisation of the thesis of inductive fallibilism. Another is the

disastrous misnaming, by Carnap, of the general theory of the probability of one proposition in relation to another as "inductive logic". See Stove [1973], Chaps. 7, 8 and
Chap. 1, pp. 21–23.

[11] See Kemp Smith [1964].

[12] See Freud, "Civilized Sexual Morality and Modern Nervousness" in Freud [1924],
II; pp. 76–99.

[13] In other words (A) holds given only that our assessments of probability are, in
Carnap's terminology, both "regular" and (with respect to individual constants) "symmetrical". This is the foundation of Carnap's proof that (his versions of) the classical
"Theorems" associated with the name of Bernoulli hold "for any regular symmetrical
confirmation-function". See Carnap ([1962], Chap. 8).

[14] The words here are my own, but the views are a summary of those expressed by
Popper in his ([1968], pp. 286–287, 296–297).

[15] Popper writes: "I have had students who thought that 'All men are mortal; Socrates
is mortal; thus Socrates is a man' was a valid inference. Yet they realised the invalidity
of this rule when I discussed with them a counter-example (arising from calling my cat
'Socrates')," ibid., p. 297. It appears then that Popper taught these students that one
can learn from a counter-example that every instance of "All F are G and x is G, so x is
F" is invalid. If so, he taught them two falsities. For G = red, F = coloured, for example,
"$P(x$ is F/All F are G and x is $G) < 1$" is false. And even if it were true, its truth would
not be learnable from a counter-example, for various reasons, among them the one
mentioned in the text: the empirical inaccessibility of the symmetry of individual
constants.

D. H. MELLOR

CHANCE AND DEGREES OF BELIEF[1]

1

Are there chances, and if so, what are they? These are still contentious questions. Like Salmon, I believe that there are chances, but I disagree with him about what they are. We do agree on some points, principally that chance is a species of objective probability, namely physical or statistical probability. Some thinkers have claimed to discern another species in this genus, namely the relational probability that inductive logic treats of. I doubt the claim: inductive probabilities are, I suspect, all descended from chances. Certainly most are, so chance is anyway the right species to study first. Whether it exhausts the genus is a question we can afford to leave open.

The first question to ask about chance is what, if anything, is wrong with a frequency account of it. There is clearly an intimate connection between chance and frequency: of heads on coin tosses, of smokers getting cancer, of radium atoms decaying in set times. There need to be good reasons for not identifying chance with frequency. Unlike Salmon, I think there are such reasons, and that sets me the task of providing an alternative account of chance, one which, amongst other things, explains how it does relate to frequency.

Frequency theories of chance have been around a long time (e.g., Venn [1866]), and their merits and defects are well known. Their overriding merit used to be that no other theory made sense at all of chance being, as it is, both empirical and objective: frequency theories had the market to themselves. Today they have competitors, a motley collection of so-called "propensity" theories, which treat chance as something like a weak disposition: of coin tosses to land heads, of smokers to get cancer, of radium atoms to decay. These theories, mine (in *The Matter of Chance*, MC for short) included, at first sold well enough on the familiar defects of Model T frequentism. But now, perhaps stimulated by the competition, Salmon and others have improved the frequency theories' specification. It is still imperfect, but so are its rivals, as Salmon [1979f] has rightly pointed out. They too need improvement and defence if they are to remain on the market. My object here, in response to Salmon's challenge, is to improve the specification of

49

Robert McLaughlin (ed.), What? Where? When? Why?, 49–68.
Copyright © 1982 by D. Reidel Publishing Company.

my own propensity model, without, I hope, adversely affecting its conceptual economy. But since the basic design remains the one I gave in MC, I shall draw on that as I go, without elaborating, amending or defending it except where it has been alleged to be defective.

2

I start with subjective probability. This is a measure of the strength of beliefs that satisfies the standard axioms of mathematical probability. Thus the strongest possible belief has probability 1, the strongest disbelief has probability 0, and the various degrees of doubt have probability values in between; if the probabilistic degree of belief is p, that of its negation is 1-p; and so on. One way of applying a measure of my belief's strength is for me to think what odds I would choose for compulsory bets on its truth in a betting situation so specified as to prevent anything else influencing the choice of odds. A so-called "Dutch Book" argument then shows that a simple function of these odds, the betting quotient, is constrained to satisfy the probability axioms: otherwise, in the situation specified, I would lose on some outcomes and win on none, which is no bet at all. When a betting quotient is so constrained, it is called coherent; I shall refer to it as a CBQ for short.

I take the Dutch Book and other arguments to show that the strength of most of our beliefs has a probability measure, and this is what I mean by subjective probability. Not everyone agrees: Kyburg [1978], for example, does not. But I have tried elsewhere [1980] to meet his objections to subjective probability, and here I shall take my riposte as read, since using probability to measure degrees of belief is not the most contentious aspect of my propensity theory. My main task will not be to defend subjective probability but to say how and why chances should constrain it. The constraint is clearly to be imposed somehow by what we know; but not just by our knowledge of non-chance facts limiting the degree of our belief through being more or less good evidence for its truth. That relation (of being more or less good evidence) is an *a priori* inductive probability if it is a probability at all, and inductive probability is not our concern. Chances, if there are any, are in the world, not just in our inductive logic: knowledge of them is part of our evidence, not just a measure of how good the evidence is. The question is, how chances are in the world, i.e., what knowledge of chance is knowledge of, and why and how such knowledge should constrain the degree of our beliefs.

On the first part of this question I have little to add to what is in MC and my [1974] article 'In Defence of Dispositions'. Most of our knowledge of chances I take to derive from knowledge of things having a kind of dispositional property which, taking the term from Popper [1957], I call 'propensity'. For example, the bias − or fairness − of a coin or of a coin-tossing gadget is a propensity, and so is the half-life of a radium atom. To know a propensity is not, however, immediately to know a chance. It is only to know what chance there would be, e.g., of heads *if* the coin were tossed, or of the atom's decay in a set time *if* it remained unbombarded for that long; i.e., what degree of belief in those outcomes could be objectively justified in those circumstances. Propensities are what makes true what Lewis [1979] calls 'history-to-chance conditionals'; only, since history does not act directly at a temporal distance, they are properties of things at the very times to which the consequents of the conditionals refer. Propensities, in other words, are not chances, but dispositions whose displays are chances; just as fragility is a disposition whose display is the fragile thing breaking. In both cases, the disposition is a property of a thing − a coin, an atom, a fragile glass − whereas the display is a property of an event − the coin being tossed, the atom being unbombarded for a stretch of time, the glass being dropped.

Few advocates of propensity distinguish it from chance in this way, unfortunately. Most just call chance 'propensity' because they take chance itself to be a disposition, namely a disposition to yield a "long run" frequency, e.g., of heads if the coin toss were endlessly repeated. But this does little more than a good frequency theory can do. In particular, it does not explain how knowledge of such a disposition should tell us how strongly to believe that heads will be the outcome of a single toss. Propensity theories of this kind mostly deserve Kneale's charge (Körner [1957], p. 80) against Popper's account: they think calling chance propensity is enough to solve the conceptual problems it presents.

Such theories prostitute the good name of propensity; they are no better than they should be, and I desire not to be associated with them. What I do desire to be associated with are theories that treat of dispositions, like colour and temperature, to affect us in ways which the dispositions explain and by which they can in turn be characterised − colours by the visual sensations we get when we look at them, temperatures by the feelings of warmth they generate in us. The best way to understand propensity is to compare it with such dispositional properties, in order to see just how and why it differs from them. Any such property would do; I take colour because it is amongst the most familiar and least technical.

3

A thing's colour may be specified by the colour of the light reflected from it, and this in turn by the subjective colour judgement which perceiving the thing by this light should produce. Similarly, a propensity may be specified by the chance which displays it, and the chance in turn by the degree of belief which perceiving it, and hence the propensity, should produce. In both cases we are specifying a disposition by its display, and the display by the mental state which should result from perceiving it and through it the disposition. The question is whether this is an intelligible and useful way of specifying an objective property and, in particular, how it can be done for propensities.

To specify a disposition in this way we must do two things. First we must identify the mental state involved independently of the disposition it will be used to specify. 'Looks green' must not just mean 'is the colour green things look'; 'seems probable' must not just mean 'is how probable probable things seem'. Otherwise the specification would be unenlighteningly circular. Second, we must say when being in this mental state amounts to knowing that something has the disposition in question. When is seeing green having knowledge of green light and thence of a green thing? When is a degree of belief knowledge of a chance and thence of a propensity?

I shall undertake these two tasks in turn, confining myself mostly to the relation between the display and the mental state, i.e., between chance and degree of belief. Once we get that right, the relation of chance to propensity will, I believe, pose no problems not adequately dealt with in MC and my [1974].

Take colour first. The mental state to be identified when I see a colour is the having of a belief with a certain content, e.g., that something is green. Now we need not identify this state completely. For instance, we need not distinguish it from believing that something else is green: it is only the colour content we are after. What we must distinguish it from is believing that something is red, or hot, or anything else but green.

We start by appealing to the peculiar visual experience people typically have when something looks green to them. There certainly is such an experience, which differs from that of seeing something red or feeling it hot. The problem is to say what it is. For we apply our colour words primarily to things, not to their looks; and it is tempting to think that 'looks green' does just mean 'looks the colour green things look'. Suppose, for example, that fresh grass is a paradigm of a green thing. Does 'green' not mean in part 'the colour of fresh grass', and 'looks green' therefore 'looks the colour fresh grass looks'? If so, specifying colours by their looks will be going in a futile circle.

Fortunately it is not so. The look of a paradigm is what we baptise, not the paradigm itself. Suppose a foreigner wants to know what colour 'green' stands for. It is no use my telling him that fresh grass is green unless he sees some. And having seen some, he can perfectly well conceive that the grass might have looked quite different, i.e., have been a quite different colour. The paradigm only has to be green in the actual world, we might say, not in all possible worlds. In other words, 'fresh grass is green' does not have to be analytic for fresh grass to be a paradigm of green; 'looks green' therefore does not in that sense mean 'looks the colour fresh grass looks' (cf. Kripke [1971]).

'Looks green' applies to whatever, in any possible world, looks the same colour as our paradigms look in this world. And whether two things look the same colour to someone is nothing more than the question whether he can tell them apart when their other properties are hidden from him. That is how we detect colour blindness, where different colours look the same, and similarly, how we detect the acuity that can sort out colours indistinguishable to the rest of us.

So far so good. We can, I believe, identify the mental state characteristic of a colour perception without disabling reliance on the actual colour of the thing perceived. But whether or not that is really so for colour, it is certainly so for propensity and chance. Here it is the strength, not the content, of a belief which has to be identified. The content can be anything which we could think to have a chance: a coin landing heads, a man getting cancer, an atom decaying within the year. In each case what distinguishes the perceived chance is only the degree of the belief. So we need not trouble to distinguish beliefs themselves, only the different degrees of any one of them.

I have alluded already to the standard method of distinguishing degrees of belief as subjective probabilities. I admitted that this has been jibbed at; but not for appealing to chances. No one thinks we need to know the real chance of an event in order to measure the strength of someone's belief in it. On the contrary, many subjectivists are positively motivated by determinism: it is just because they suspect there are no objective chances that they invoke subjective probabilities to provide a surrogate subject matter for statistical science. The surrogate is indeed inadequate, but their view is not inconsistent, nor is it epistemically problematic: everyone agrees that we can detect degrees of belief just as well in a deterministic world as in a chancy one.

The problem with chance is not identifying the mental state involved; it is saying when being in that state amounts to having knowledge. When is having a degree of belief knowing a chance? Here the case of colour presents the easier problem; and since its solution will set us up to tackle chance, I will again take colour first.

4

When is seeing something green knowing that it is? The stock answer is: when it really is green and our seeing it that way is somehow justified. The second part of this answer I shall follow Ramsey [1929] in taking to be a matter of reliable causation. In a good light, a thing's being green will reliably cause me to see it green when I look at it, and its not being so would reliably cause me to see it otherwise. In those circumstances, my seeing it green amounts to my knowing that it is.

The problem this answer presents here is that it appeals to the thing's really being green. This fact is needed both to make my belief true and to justify it by being a reliable cause of it. But I am trying to specify this very fact in terms of the thing looking green; and it has to be shown how that can be done without trivialising this answer. There must be more to being green than looking green, or the answer will put no objective constraint at all on our knowledge of colour. And what more there is must be statable without appeal to the colour of things.

We can start by saying that a thing is green when it looks green to all or to most people. It should be clear from Section 3 that there is no vicious circularity in saying this and, though it is not quite right, it is a good first approximation; and seeing what it entails will enable us to improve on it. What makes something seen by reflected light look green to most people is that it reflects light predominantly of certain wavelengths, and these make most people see things green. Indeed light of these wavelengths makes everyone see things green who is not colour blind nor has something else wrong with his eyes. In other words, there are laws linking the properties of things, of light, and of our eyes and brains, to the colour things look to us. These laws are what make certain things such that all or most people would see them green; more precisely, such that all normally-sighted people would see them green under normal white light. And that is what it is for something really to be green.

Note that this specification of greenness is not circular. Apart from how things look, it mentions no colours, only such properties as reflecting light of certain wavelengths. Nor is it as unduly democratic as it seems. Normal sight is specified physiologically, not by Gallup poll; and normal white light is specified as a certain spread of wave-lengths. As with our paradigmatically green grass, while we naturally choose a specification which fits our actual lighting and eyesight, that fact is not built into the meaning of the word 'green'. It makes perfect sense to imagine radiation so affecting our eyes that we would all come to see green things red.

Nevertheless, being green is specified in terms of how things actually look to us. There remains an essential sense in which a blind man does not know what green is, because he does not know what it looks like.

Still, in order to be green, a thing must satisfy certain laws. Colour is therefore not a purely phenomenal property; indeed, like Hesse ([1974], Ch. 1), I suppose that nothing is. It is, in fact, obvious that even our everyday observation of colour is not just a matter of recording visual sensations. We have digested enough of the relevant laws to enable us to see the same objective colour through very varied visual experiences by allowing, quite unconsciously, for other factors like shadows and reflections. In the same way, we can likewise see varied colours via identical visual experiences. A dimly lit white surface, for example, will give us the same visual sensation as a brightly lit grey one; yet we can tell the colour difference, by seeing how differently they are lit. And even with all this unconscious correction, we know our colour judgement is not infallible. Even when well-informed, it can still be overruled by evidence that the lighting is odd or our eyesight defective.

We can use our laws to anticipate our visual sensations, as well as to check and to correct them. Thus we can know — or even, in paint manufacture, make — something to be green which no one has ever looked at. Our blind man may not, in one sense, know what green is; but by knowing its laws, he can say what it is, case by case. He can say, for example, that what makes monochromatic light green is its wavelength lying between such-and-such Ångstrom units. In other words, these laws supply truth conditions for statements about the colour of things: 'monochromatic light is green' is true just in case its wavelength lies in the specified range.

Now we who can see well enough to discover and test the laws that govern colours, need not be able to formulate them all in order to judge the colours of things. But the existence of some such laws, and of the truth conditions they entail, is nonetheless our warrant for admitting colour to be an objective as well as a phenomenal feature of the world. It is what entitles us to propose seriously the seemingly trite truth condition: 'X is green' is true if and only if X is green. The warrant assures us that there really are such facts as things being green, which can therefore determine the truth of statements and beliefs, whether about colour or about other matters. But then we can in particular use these facts to give the stock answer to our original question. Seeing something green amounts to knowing that it is just when it really is green and our seeing it so has been reliably caused, e.g., by the fact that it is so.

5

In saying when it is that seeing green is knowing it, I have appealed to the content of the belief, namely that something is green. The content is what specifies the fact which, if it obtains, both makes the belief true and provides in perception a reliable cause of it. But this sort of account appears not to work for knowledge of chance, because chance is supposed to constrain the strength of a belief, not its content. Knowing the chance of an event is having a justified degree of belief in its occurrence. The chance does not figure in the content of this belief; nor does its content specify the fact we need in order to justify its degree. Neither the event's occurrence nor its nonoccurrence justifies believing in it to degree 0.5 say, rather than to degree 0.3 or 0.6. So what, if anything, does?

Frequency is the obvious candidate: how often such an event occurs in suitably similar circumstances. At first sight this seems to enable us after all to account for knowledge of chance on the standard model: knowledge of chance is true belief, causally justified, in the relevant frequency. In fact, this answer will not do. Suppose there is only one such event: e.g., a coin is tossed once only. The frequency of heads will be 1 or 0, but this clearly does not make 1 or 0 the chance of heads on that toss. And even supposing we do have a credible frequency, say fifty heads in a hundred tosses, still this would not justify that degree of the relevant belief. For if this frequency is a chance at all, it must be the chance of heads on each of those hundred tosses, i.e., it must prescribe the degree of belief we should have in heads being the result of any one of them. Now the whole point of such a chance is to be knowable in advance. We must be able to justify belief of degree 0.5 before knowledge of the actual result enables us to justify belief of degree 1 to 0. But the frequency depends on the actual results; it is not causally available in advance. The frequency of heads in a hundred tosses could not be a justifying cause of my degree of belief in heads on the first toss in the way a thing's being green can be a justifying cause of my seeing it green.

So actual frequency will not serve. The right one usually does not exist, and is causally useless when it does exist. The frequentist's next resort is hypothetical long-run frequencies: how often the coin *would* land heads if only it were endlessly tossed in the same way. But this will not do either. Coins are never endlessly tossed: hypothetical frequencies are fictions, not facts, and fictions can neither cause beliefs nor make them true. In particular, there is no one frequency of heads among merely possible tosses of a coin. Tosses with other frequencies of heads are just as possible as those with the frequency 0.5 which is supposed to justify that degree of belief.

It must, if anything, be a fact about the actual toss that, were it to be endlessly repeated, 0.5 would be the long-run frequency of heads. This fact is the disposition postulated by the propensity theories I disowned in Section 2. The disposition might indeed exist, and it might even cause the degree of belief it is supposed to justify. But how is it supposed to justify it? What has a disposition to produce a long-run frequency of heads on other, mostly nonexistent, tosses to do with my prospects of getting heads on this actual toss? It is no use doing what Hacking does ([1965], p. 135) with his "frequency principle", namely in effect defining the concept of justification or support to be such that this disposition supplies it. That just provokes the question: why should I adopt for this toss the degree of belief that is justified in that sense? And to that question I know of no sufficient answer. We cannot get to justified degrees of belief starting from frequencies, actual or hypothetical, or from dispositions to produce them. The only way to start, as with colour, is by specifying the fact by the degree of belief it is supposed to justify. And if we do that, the frequencies, as we shall see, will take care of themselves.

6

So now I must say how chance can be specified as a fact justifying degrees of belief, if not by making beliefs about frequencies true and justifying them. I also have to show that chance, when it is so specified, will relate to frequency in the way we know it does; and it is, in fact, simpler to tackle this problem first.

I start therefore by assuming that there are chances, waiving for the moment the question of what they are and how they work. The immediate problem is to relate them to frequencies. Now I believe I solved this problem in MC (Chapter 8), but Salmon [1979f] thinks not. So, without repeating the solution in every detail, I need to sketch its salient features and defend it briefly against Salmon's attack.

Suppose a degree p of belief in a coin landing heads is justified by knowing that p is its chance of doing so. To get what follows from this supposition about the frequency of heads on repeated tosses, I appeal to the laws of large numbers. By these it follows in particular that as high a degree p^* of belief as we like, short of 1, is justified in any frequency proposition F of the following form: in enough such tosses, the frequency of heads would be within δ of p, where δ is any positive real number, however small.

This very high degree p^* of belief in the proposition F looks as if it needs another chance to justify it; and if that were so, there would be an arguably vicious regress. But p^* does not need a chance to justify it, as the following argument shows.

In Chapters 1 and 2 of MC, I argued that a degree of belief increasing towards 1 must turn into full belief before it gets there. Salmon ([1979], p. 187) himself remarks that no one's beliefs would make him risk unlimited loss for a penny gain, i.e., that none at least of our contingent beliefs is actually of degree 1.[2] For a belief to be justified, therefore, its justified degree sufficiently close to 1. This being so, it follows that as the justified degree need not be 1; it need only be a degree sufficiently close to 1. This being so, it follows that as the justified degree of belief tends to 1, it will turn into justified belief somewhere before it gets there. Where it does so will depend on context, but that is immaterial here, since the laws of large numbers can get the justified degree of our belief in any F as close as any context could conceivably require. So whatever the context, our assumption justifies us in simply believing every F, and hence in believing that, in a sufficiently long run, the frequency of heads would come indefinitely close to the chance p. And what makes that belief true is just that the toss has the disposition, to produce long-run frequencies, to which I alluded in the last section.

the corresponding hypothetical long-run frequencies. Specifying chance in this way will not sever any of its proper frequency connotations. But what is the specification, and how does it work?

7

I must first say what it is for a degree of belief to be justified by a fact about an event. The sense of justification we require has to be factual rather than, say, moral: an outcome of the coin toss is to some degree to be expected, not to be approved or deplored. Yet justification here is not a matter of making a belief true, since truth applies to the content of a belief, not to its strength. There needs to be some other mode of justification by facts which relates specifically to the degree of a belief rather than to its content.

To see what this mode of justification is, consider again the connection between betting and degree of belief mentioned in Section 2. I argued in MC that people's choices of coherent betting quotients (CBQs) show how strong they think their beliefs are, provided they suppose the betting situation to be restricted in specified ways in order to exclude any effect on their choice of attitudes, other than the belief whose strength is to be measured. I then used the entailment just established to show that only at a CBQ = p can I know that I would break even in a long enough run of bets on coins landing heads when their chance of doing so is always p. Now under the restrictions needed to make my CBQs measure my belief, I must suppose myself compelled to bet without control either of the stake or of the direction of the bet. In that situation breaking even is the best result I could possibly hope to know of. So I have a plain gambling rationale for choosing this CBQ in that situation and, hence, for having the degree of belief which, in that situation, this CBQ measures.

We need something like this argument to show in what sense a chance can be said to justify a degree of belief. I have therefore to meet an objection which Salmon ([1979f], pp. 188–192) has brought against it. The objection is to my definition of breaking even. In MC, I defined breaking even on a run of bets as having a net loss less than the least unit of the currency, on a fixed stake divided equally among however many bets there are in the run. Salmon on the other hand, supposes a fixed stake on each bet, and defines breaking even as being no worse off after all the bets in the run. Betting Salmon's way, there is no CBQ at which I can know I would break even: the net loss is as unlimited as the number of bets – not surprisingly, since the total stake is likewise unlimited. But that is not just putting one's shirt on, as Salmon has it: it is putting on an indefinitely extensive wardrobe that no one has. In real life, we have no more than a limited amount of goods, however large, to wager with; and breaking even is, I still maintain, losing less than whatever fraction of that equals the smallest unit of the currency.

However, the core of the argument would survive even in a casino lucky enough to attract Salmon's unlimited patronage. Its patron could indeed never know that he would break even; but he could know how to lose his wardrobe as slowly as possible, namely at a CBQ = p. And as that is the best knowledge he can have in the circumstances, p is still the uniquely justified degree of his belief.

8

For me to think there is a chance of heads is thus for me to have a degree of belief in heads and to think it justified in the above sense by some fact about the coin toss. If I am right, the fact will be that the toss has some property, such that at a CBQ contained in the shortest interval which measures my belief, I can know I would eventually break even on bets on heads resulting from tosses all of which have this property.

The measure of the property, namely the chance of heads, is the degree of belief which it justifies in this fashion. Like colour, chance is being specified by how it seems. A man incapable of doubt, i.e., whose beliefs are all of degree very close to 1 or 0, does not know what chance is, in precisely the sense in which a blind man does not know what green is. He is never in the mental state characteristic of one who perceives that property.

But a man without doubts can still say what chances are, case by case, just as a blind man can say what colours are. Neither man can say it *a priori*: nothing *a priori* prescribes how alike in other ways things or events have to be to share the same colour or chance. Laws of nature are what prescribe that, as we have already seen in the case of colour. In the case of chance, the laws in question are statistical laws. A statistical law is one which says that having some other specified property suffices to fix a chance. The laws of radioactive decay, for example, specify the nuclear structure of radium to be such a property. The law says that all atoms with that structure are such that, if I were to bet often enough at a CBQ = 0.5 on them decaying in the next 1622 years, I can know I would break even; so that they justify my having that degree of belief in any one of them doing so.

In other words, laws supply truth conditions for statements about chance, just as they do for statements about colour. 'The chance of this atom decaying within 1622 years is 0.5' is true just in case the atom's nuclear structure is that of radium. Now those of us who are capable of doubt, and thus of perceiving chances, need not know these detailed truth conditions in order to detect a chance. We can detect a chance behind sufficiently suggestive frequency data, for example, or from evidence of symmetry in a coin, without knowing the particular laws which govern it. But the existence of such laws, and of the truth conditions they entail, is, as with colour, our warrant for taking chance to be an objective feature of the world. To say there is a chance is to claim such a warrant. We may not know the statistical law governing a particular coin toss, but unless there is one, the toss has no objective chance of landing heads.

9

'The chance of X is p' is true if and only if the chance of X is p. This is not as trite a truth condition as it seems. It claims there are such facts as X having a chance, and that is so only if X is subject to statistical law. However, although this truth condition is not trite, it can be misleading in a way the corresponding one for 'X is green' is not. It can easily suggest that the role of chance facts is to make true a belief with a particular content, namely a belief about a chance. But that, we have seen, is not so. Its role is objectively to justify a particular degree of a belief about something else. There is an important sense in which statistical science has no subject matter of its own. The subject matter of a statistical science is just that of its nonstatistical counterpart.

This is why it has proved so hard to frame an acceptable account of objective chance. People naturally feel that, if chance is objective, it must make true beliefs with some characteristic content. Frequency theory's appeal is precisely that it offers such a content, namely facts about frequencies. Conversely, those who have seen that making beliefs true is not what chance is for, are naturally disposed to deny its objectivity; hence the popularity of subjectivist accounts of it. Where both parties err is in supposing that the only objectifying job facts can have to do is making beliefs true.

But this is not so, as we may see in ethics. An objectivist there will naturally give truth conditions, such as: 'X is good' is true if and only if x is good. This truth condition, like that for chance, is not trite: to give it, the objectivist has to think there are such facts as X being good, and that X's having some other properties would make that fact obtain. But he is not thereby committed to thinking that this fact's role is to make true a belief whose content is that X is good. Obviously not: its role is objectively to justify a mental state quite different from believing, namely approving of X. The content of all the relevant *beliefs* about X is entirely non-moral. Morality, like chance, has no subject matter of its own. It may be objective nonetheless; beliefs are not the only mental states capable of objective justification. Failure to see this underlies two views in ethics which correspond closely to frequency and subjective views of chance. One view tries to provide a distinctive content for the belief that X is good; e.g., that X promotes human happiness. The other view sees that this misses the whole point of morality (since it remains an open question whether one should approve of promoting human happiness or anything else), and concludes that there is no objective goodness at all.

Properly to recognise both objective goodness and objective chance, we need to extend our conception of facts as suppliers only of true beliefs. The extension is less for chance than it would need to be for goodness, since chance at least has to do with belief, just as truth does. Indeed I suspect that in the end truth will prove to be no more than an extreme case of chance. To call a belief true is, after all, to say that objective justification exists for believing it to a degree very close to 1. An adequate and comprehensive account of how degrees of belief can be justified should be able to make a separate theory of truth redundant.

10

I am not, however, trying yet to sack all truth-making facts, only to employ some chance facts as well. Up to now, though, truth-makers have run a virtually closed shop in the facts business, and I foresee considerable opposition to these chancy new recruits. And they really necessary? Is there really objective work to be done which only they can do?

The work chance facts do is justifying degrees of belief. Now, according to Bayesian decision theory, degrees of belief combine in a familiar way with degrees of desire to justify choices of action. For example: I have to go out, don't like carrying an umbrella needlessly, but also don't like getting wet. So whether I should take one depends on the degree of my belief that it will rain. Pure subjectivists would maintain that there is no need to justify the degrees of belief which in turn justify courses of action, and so there is no need for chances. But a subjectivism which is that pure is not really credible. A degree 0.99 of belief in imminent heavy snow might well justify travelling with a snowplough; but not in the Sahara in July, because there that degree of that belief would be objectively absurd.

Perhaps, however, there does not need to be a chance of snow, only an inductive probability of it relative to truth-making facts about temperature and humidity? Not so. Bayesian decision theory does need chances, as Jeffrey [1980] has remarked. Suppose I am a smoker deciding whether to give up the habit because of my fear of cancer. For me, smoking "dominates" not smoking, i.e., I shall prefer to smoke whether I have cancer or not. But I also know I should very much prefer not to have cancer, and I think cancer much more probable if I smoke. In short, the degrees of my relevant beliefs and desires make the theory tell me to give up smoking. And so I should, but only if what justifies the degrees of my conditional beliefs are propensities, not merely inductive probabilities. Since I prefer to smoke in any case, I would be a fool to quit if my smoking were merely better evidence than my not smoking for the hypothesis that, whether I smoke or not, I shall get cancer. Quitting can only be justified if it is an action which will cause a change in my prospects, namely a reduction in the probability of my getting the disease. But a probability which has causes is a part of the physical world, not merely part of an inductive logic; that is, as I remarked in Section 2, a chance. In other words, inductive probabilities are not enough to make sense of the prescriptions of Bayesian decision theory. The work objective probability has to do there can only be done by chance.

Chances of course are not only effects, as in the case of smoking and cancer. They are causes too, and are invoked by statistical theories to explain phenomena. Statistical mechanics, for example, invokes a chance distribution over the positions and momenta of gas particles in order to explain the gas laws. The exploding of an atomic bomb is the effect of increasing the chance of it recapturing its own fission products. Chance facts figure as prominently in the causal history of the world as do the facts which make beliefs true.

Yet the need to employ chance facts might still be resisted. We have seen that statistical laws give apparently non-chance truth conditions for chance statements. To every chance there seems to correspond a truth-making fact: e.g., that a radium atom has such-and-such a nuclear structure. Do not these non-chance facts suffice to justify degrees of belief, and to fill the role of chances as causes and effects, as well as providing inductive probabilities? Perhaps the chance of heads is not itself a property of a coin toss, but merely a way of referring to whatever non-chance property actually justifies a particular degree of belief in that outcome of it.

This view is the statistical analogue of a longstanding view of dispositions. Fragility, it has been held, is not itself a property of a fragile glass. Rather it is a way of referring to whatever non-dispositional property, such as the glass's molecular structure, is causally responsible for making the glass break when dropped. It will then be a law of nature that whatever has this property is fragile; just as it is a law that whatever has the nuclear structure of the radium atom has a fifty-fifty chance of decaying in 1622 years. In each case it is the law which supplies the truth conditions; but given the law, the only facts we need admit are that the glass has such-and-such a molecular structure, and the atom such-and-such a nuclear one. Neither dispositions nor chances need to figure as constituents of facts in their own right.

I have attacked this view of dispositions elsewhere [1974]; basically because the real properties which are supposed to provide truth conditions for fragility are as dispositional as it is. There are simply not enough non-dispositional properties of things to go round. It takes other dispositional facts to account for glasses being fragile, and for changes in fragility being causes and effects. The other facts may be microscopic, but that does not make them non-dispositional. And so it is with chance. There are not enough non-chance properties of events to go round. Nuclear physics is irreducibly statistical; the nuclear structure of radium is not in reality a non-chance fact at all. For an atom to have that structure is for there to be certain chances of nuclear events occurring within it. It takes chance facts, not just truth-making facts, to account for radium's objective radioactivity. The facts may be submicroscopic; but they are no less chancy for that.

11

I conclude that chances are indispensable ingredients in any world governed by statistical laws. Such a world need not, however, be indeterministic, as I thought when I wrote MC. That was an error, as Salmon [1979f] and others have convinced me; and I am pleased to take this occasion to recant. But I need then to show how my chances can be made compatible with determinism, and to rebut my earlier arguments to the contrary.

Consider first a coin landing heads with chance 0.5. The chance justifies that degree of belief in heads; but the fact of heads also justifies a degree close to 1 in the same belief. How can two different degrees of the same belief both be justified? The answer is that the two justified degrees are not actually degrees of the same belief. There are two beliefs: the past tense belief that the coin has already landed heads, and the future tense one that it will do so. The two beliefs are both made true, if they are, by the same fact, namely the coin landing heads, but they are not the same mental state. For one thing, they relate differently to action (see Perry [1979]): there are circumstances in which believing a coin will land heads would make me behave quite differently from the way I would behave if I believed it had already done so. For another, although the same fact makes both true, it makes them true at different times. The past tense belief can only be true after the coin has landed, the future one only true beforehand. But since causes always precede their effects, the coin landing heads can only be a justifying cause of the past tense belief while it is true. A future tense belief of which it could be a cause would have to be about another, later event. I cannot see that a coin *will* land heads, only that it *has* done so.

So only the past tense belief is justified to a degree close to 1 by the coin landing heads; the future tense one is not. The coin landing heads does not therefore prevent the future tense belief being justified merely to degree 0.5

by facts, about the propensities of the coin and tossing device, which are causally available before the toss. I can see beforehand that a coin is fair, i.e., is such that, if tossed in a standard way, the chance of it landing heads would be 0.5. And I can know it will be tossed that way because I intend to toss it, and I can. Hence I know there is a chance 0.5 of the coin landing heads on this future toss, i.e., that future tense belief is justified to degree 0.5. That I know the corresponding past tense belief will subsequently be justified to a degree close to either 1 or 0 does not detract from this knowledge at all. On the contrary, without knowing that, I could not envisage settling a bet on the coin landing heads; and I can make no sense of justifying any degree of any belief without being able to settle bets.

Generally therefore, it is only future tense beliefs which are justified to degrees remote from 1 and 0. (Betting on past events can of course occur, but it always depends on the future disclosure of crucial evidence not yet available to us.) But it is the degrees of future tense beliefs which matter to decision theory, as the example of Section 10 shows. If smoking makes it probable that I will get cancer, that can make it sensible for me to stop smoking. But not if it only makes it probable that I already have cancer, i.e., that it justifies only a high degree of the past tense belief. Just because causes always precede their effects, such a past tense probability could not be an effect of my present smoking. It could only be an inductive probability, not a chance, and my decision therefore should not turn on it.

Similarly with coin tossing. If it matters to me whether a coin lands heads, then all I can hope to affect is the probability that it will do so, not the probability that it has already done so. If I toss it one way, the chance of heads is p_1, if I toss it another, the chance of heads is p_2: these are propensities of tossing devices. Knowing them is what justifies the degrees of the future tense conditional beliefs which I need in order to decide which way to toss the coin. So long as propensities and the chances that display them can justify intermediate degrees of future tense beliefs, they do all that decision theory needs. That, so far as past tense beliefs are concerned, they are overridden by the actual results of the tosses is neither here nor there.

But now suppose the result of the toss is, in fact, determined by earlier facts, albeit ones not readily observed. Do they not justify a degree close to 1 or 0 in the future tense belief? And if so, how can a degree 0.5 be justified in the same future tense belief? For recall the betting situation I appealed to in giving a relevant sense to justification: in it, the believer's opponent, not himself, fixes both the stake and the direction of the bet, i.e., who wins if the coin lands heads. If the result is determined, and thus perceptible, in advance, anyone betting at a CBQ of 0.5 would surely be cleaned out; and wherein then lies the justification of that degree of belief?

These were the considerations which persuaded me in MC that chance was incompatible with determinism; since I no longer believe that, I have to say what is wrong with them.

The fact is that our theory of knowledge needs a causal account of justification, whether of beliefs about colour or about chance, only because our conceptual and perceptual abilities are limited. Were that not so, we should not need to distinguish knowledge from true belief in terms of justification at all. In the case of God, for example, we need no such distinction. What God knows, He knows immediately, not through the mediation of senses such as we have. He does not acquire His knowledge by sensory mechanisms of possibly variable reliability, of whose actual reliability we must therefore be satisfied before conceding that a belief is more than coincidentally true.

But we are not gods. Our knowledge comes to us by fallible means, and we need to distinguish it from coincidentally true belief. That is why we need a theory of justification for belief, a theory therefore of our limited perceptual and conceptual abilities. Since it is our theory, our conceptual limitations will be built into it; what it will deal with explicitly, and what anyway matter here, are our perceptual limitations. These need not be specified precisely, but they must be presumed. In particular, if our gambler cannot tell in advance that a coin will land heads, we must suppose that his opponent cannot tell that either. We are concerned with bets amongst ourselves, not bets with the All Seeing – a self-evidently foolish pastime at any CBQ.

So intermediate degrees of future tense belief can be justified, and chances can exist, even in a deterministic world. Determinism therefore I now take merely (but still falsely) to assert that every event instantiates the consequent of a non-statistical causal law, not that it does not also instantiate a statistical one.

12

By the same token, I can make sense of an event instantiating more than one statistical law, attaching different chances to the same result. Most events do not, but some do. The famous example is Laplace's biased coin, of which the bias is unknown. It seems that on the first toss, the right CBQ for a bet on heads is 0.5, even though the chance is known to be something else; how can that be on my account? The fact may be that there are two chances of heads: 0.5 as well as the chance that displays the coin's bias. There is nothing paradoxical in one outcome of an event having two chances, even if it is uncommon. All it means is that two degrees of the same belief could both be objectively justified; and that can be so even if no one could actually have them both at once. It can be so, because the CBQs that measure them enable knowledge of breaking even in two different betting situations, depending on what counts as repeating the toss. In one, the same biased coin is repeatedly tossed, and it is at some CBQ other than 0.5 that one can know one would eventually break even. In the other, the other bets would be on other biased coins, of which we suppose half would in the long run be biased towards heads. The fact we suppose then, is that biasing mechanisms have equal propensities to bias coins towards heads and tails; and it is our supposed knowledge of this propensity which justifies the degree 0.5 of our belief in heads on the first toss. (For the detailed proof of this, see MC, pp. 129–136.)

But which degree of belief should we actually have on the toss of such a coin? Well, either 0.5 or the degree that reflects the bias of the coin would do. But the value 0.5 is known, insofar as we know that biasing processes do not distinguish heads from tails. Whereas the other value is unknown, because our perceptual limitations, Laplace supposes, prevent us perceiving the actual bias of the coin. So it is no more practicable to recommend that other value than it is to recommend full belief in whichever side will actually land heads. And that is why, *faute de mieux*, one should bet at evens in this case.

13

Chances may thus be said to be relative to our perceptual limitations. That does not make them relative to all descriptions of events, nor to the actual frequencies which those descriptions determine. A coin toss may be one of seventy made within the shadow of St Paul's, of which twenty-eight land heads. It does not follow that there is this or any other chance of landing heads that is common to all these tosses. There is no such chance: no statistical law applies to tosses under the description 'thrown within the shadow of St Paul's', and there is no CBQ at which we could know we would eventually break even on repeated bets on such tosses landing heads. Yet any one of these actual tosses may, under some other description, be governed by one or more statistical laws; and if so, one or more degrees of belief in its outcome being heads could be justified in beings of suitably limited perception.

I have specified chances by the degrees of belief they can justify in beings of our causally limited perceptual ability. In the same way, I have specified the colour of electromagnetic radiation of various wavelengths by colour judgements they can justify in beings of our causally limited eyesight. That is a perfectly good way of specifying objective properties; ultimately indeed the only way, since science must deal in phenomena, not noumena, if its epistemology is not to be merely magical (see Mellor [1973]). But whether there are beings like us or not, the properties so specified remain. The effects of radiation and of chance are, we have seen, not limited to their effects on us. Maybe the All-Seeing should have no degrees of belief other than 1 and 0; but he can still know the world to be such that we should. That is as much a fact about it as any other; and so therefore is chance.

Cambridge University

NOTES

[1] This paper is my reply to Salmon's discussion review [1979f] of *The Matter of Chance* and other propensity theories. Although we still do not agree, I owe much to Salmon's scrupulous critique, which I try here to answer constructively. My thinking on these topics has also been especially stimulated recently by the work of Blackburn [1980], Jeffrey [1980], Kyburg [1978], Levi [1977], Lewis [1979] and Skyrms [1980]. I thought it would be clearer for me to develop my own ideas here without much cross-reference to points of agreement and difference with others; but my debt to these works should be acknowledged. So should my debt to the Radcliffe Trust, for awarding me a Radcliffe Fellowship, during my tenure of which this work has been done.

[2] But what, Salmon then asks, if I believe a chance to be 1, as I clearly can and often do? Does that not destroy a correspondence between chance and degree of belief which my theory requires? Not at all. If none of my contingent beliefs is of degree 1, then in particular my belief that the chance of an event e is 1 will not have that degree, nor therefore will my belief that e will occur. For example, suppose I am convinced (to a degree very much closer to 1) that the chance of e is either 1 or 0.5, my degrees of belief in these chances being very close respectively to 0.99 and 0.01. Suppose also that in the circumstances this belief of degree 0.99 is strong enough to count as belief, i.e., no increase in its strength would affect any of my actions. Then I do not believe the chance of e is 1. But it does not follow that my degree of belief in e occurring is 1. On the contrary, given these suppositions, it follows that my degree of belief in e is very close to 0.995: enough therefore to constitute believing that e occurs, but not believing it to degree 1.

INVENTION AND APPRAISAL[1]

0. INTRODUCTORY ACKNOWLEDGEMENT

Among the many important contributions made by Wesley Salmon to the explication of the logic of hypothesis-appraisal has been his insistence that considerations additional to the inductive relations between evidence-statements and hypothesis may bear upon the probability of the latter. In particular, he has urged that the *prior* (to test) probability of a hypothesis *H*, as well as the *likelihoods* of *H* and not-*H*, must be taken into account in assessing the *posterior* (to test) probability value of *H*. Salmon has argued that the relationships among these several probability-values are captured by a particular interpretation of Bayes' Theorem – a theorem of the formal probability calculus (Salmon [1967a], [1970c]). If this proposal is adopted, the problem arises of how to assign a value to the prior probability of *H*. Since no frequencies directly relevant to *H* will be available prior to initial testing, Salmon, who adopts a frequency interpretation of probability, must appeal to *other* considerations in estimating this prior probability value ([1967a], pp. 124 ff). These are what he calls "plausibility considerations" for *H*. As well as formal and pragmatic criteria, they include material criteria, such as considerations of simplicity and symmetry. From plausibility considerations of these various types, a scientist would be enabled to judge the *plausibility* of his/her hypothesis prior to test, and to estimate a *prior probability* value for insertion in Salmon's Bayesian schema (ibid., pp. 115 ff, and my pp. 76–8 below). Even if, in many cases, the prior probability of a hypothesis is difficult to quantify precisely, Salmon's recognition that this factor plays an important role in the appraisal of the hypothesis has always seemed to me to be a valuable insight.

In what follows, I aim to extend this insight into the context of *invention* of hypotheses by arguing that considerations similar (and often identical) to Salmon's "plausibility considerations" may play a significant role in invention. Such "guiding considerations" as I shall call them, may enter into *advancement* arguments for *H* in the context of invention; and often the same or similar considerations (e.g., of simplicity, or analogy with a model) will serve also in *enhancement* arguments for *H* in the context of appraisal. In

69

Robert McLaughlin (ed.), What? Where? When? Why?, 69–100.

the account which I shall develop, a rationally-invented hypothesis normally will enter the context of appraisal accompanied by some considerations which bear upon its plausibility. Thus, Salmon's analysis of appraisal itself suggests, and can be integrated with, a rational reconstruction of the invention of plausible hypotheses. My present aim is to demonstrate the legitimacy, rationale and feasibility of such a rational reconstruction — as against the received view that this enterprise is illicit, idle or impossible.

1. PRELUDE: "LOGIC OF DISCOVERY"?

1.1. *Confusions*

In ordinary parlance, a physicist may speak of the "discovery" of new elements, by means of such sophisticated equipment as particle accelerators and bubble chambers; in a similar way, one refers to the "discovery" of penicillin, or of quasars. To speak thus is to use 'discovery' in the sense of first encountering something in nature, either as a result of a deliberate search (normally guided by theoretical expectations), or accidentally, as was the discovery of X-radiation (Whittaker [1960], pp. 357 ff). This is a well-established sense of 'discovery'. However, it is to be contrasted with two other senses of the term, which have been widely used by philosophers of science in particular.

Reichenbach carefully distinguished what he called the 'context of discovery' from the 'context of justification' in science.[2] In his usage, discovery — and justification — have to do, not with objects or properties in or of the world, but with *hypotheses* — that is, propositions describing the world or purporting to do so. According to Reichenbach, scientists proceed by first "hitting upon" a hypothesis, usually after some intellectual wrestling with a problem; and second, testing it by means of an experiment or set of observations, with the aim of "justifying" or confirming it. This "hitting upon" a hypothesis was what he meant by 'discovering' it. Reichenbach further argued that the items comprising the context of discovery were of no concern to philosophy, since they were not amenable to rational reconstruction — i.e., there could be no "logic of discovery"; although he did concede that the discovery of a hypothesis might properly be studied by empirical scientists, such as psychologists. In contrast, it was the rational reconstruction of the context of justification that was the central concern of philosophy of science, on this view. Despite its odd consequence that a hypothesis could be "discovered" and later turn out to be false (e.g., the phlogiston hypothesis), Reichenbach's notion of discovery has been highly influential for four decades.

The ambiguity between "discovery of entities" and "discovery of hypotheses" was compounded by Popper [1959]. The original German edition of his *Logik der Forschung* had been published in 1934. The English edition was not published until a quarter-century later, under the title *The Logic of Scientific Discovery*. Popper chose to translate the German 'Forschung' in the title as 'Scientific Discovery', whereas a more common English rendering of the term would be 'scientific research' or 'scientific investigation'. This latter was clearly Popper's intended sense of 'Forschung', for the book, and many of his other writings, were concerned with the evaluation of hypotheses *after* they had been "discovered" in Reichenbach's sense of the term. The choice of the English title of his book is doubly surprising when one notes that in his opening chapter Popper writes: "The initial stage, the act of conceiving or inventing a theory, seems to me neither to call for logical analysis nor to be susceptible of it" ([1959], p. 31). Popper was thus in complete agreement with Reichenbach on the impossibility of logically reconstructing the process of "hitting upon" a hypothesis, despite their major disagreement on other matters, such as the role of induction in science.

The confusing usage of 'discovery' as meaning *research* or hypothesis-*evaluation* (which for Popper involved attempted falsification, in contrast to Reichenbach's attempted justification) was continued more recently by Lakatos [1978a, 1978b]. In these writings he persistently uses the phrase 'logic of discovery' when he means a logic of *appraisal* of hypotheses and theories. Thus for Lakatos the words 'logic of discovery' mean precisely the opposite of what they meant for Reichenbach. Lakatos seems to have been quite insensitive to this linguistic confusion.

1.2. *Definitions*

I suggest the following terminological conventions to avoid the confusions just noted. First, 'discovery' should be confined to the sense of 'initially encountering in nature' some object or property. Second, what Reichenbach called the 'discovery' of hypotheses – i.e., "hitting upon" them – should be called the *invention* of hypotheses. This term has the merit that no oddity is involved in speaking of a hypothesis being invented and later falsified. Third, what Popper and Lakatos meant by 'discovery' should be called the *appraisal* of hypotheses. This term has the virtue of neutrality as between falsifying (Popper) and justifying (Reichenbach) a hypothesis. Then I can speak of the context (and logic) of invention, and the context (and logic) of appraisal, without the ambiguities and methodological commitments deriving from Reichenbach's original unfortunate pair of terms.

In this discussion, I shall use the term 'logic' as an ellipsis for 'logical explication'; and I shall take it that the class of logically-good arguments includes correct inductive inferences[3] as well as valid deductions. Armed with this terminology, I can now address the central question concerning the "logic of invention" — namely, can there be one?

2. ISSUE: LOGIC OF INVENTION?

2.0. *Intuitions*

Intuitions seem opposed as to whether invention may be a rational procedure. On the one hand, there is a widespread view that creativity is somehow unconscious and (thereby?) non-rational, involving "flashes of inspiration", "leaps into the unknown" or "the Eureka syndrome". Popper ([1959], p. 32) cites Bergson and Einstein (the latter somewhat unfairly, I think) in support of this view. If this is so — if great concepts and conjectures really do come "out of the blue" — then it seems to follow that no logical explication or rational reconstruction of the invention of a new idea or hypothesis is possible.

On the other hand, this seems to introduce a note of *in principle* obscurity into the investigation of nature, and against this sort of mystification some philosophers intuitively rebel. To set *a priori* limits upon rational reconstruction seems opposed to the empiricist spirit of contemporary science — seems rather a step into the past, back to magic, mystery, the occult and the forever-unknowable. Added to this is a firm impression that many important scientific ideas have been arrived-at, not "out of the blue" or within a cognitive vacuum, but rather as the end-result of intense, goal-oriented intellectual labour. To insist that such intellection is ultimately non-rational seems highly implausible.

2.1. *Received View*

Despite the latter considerations, an apparent majority of empiricist-oriented philosophers since the 1930s has insisted that there can be no logic of invention, and that the study of invention is no business of philosophy — although it may be appropriate for such empirical disciplines as psychology or history. On this view, to attempt to explicate a logic of discovery is to commit the category-mistake of conflating logic with psychology ("psychologism"). Advocates of this orthodox or received view include Popper [1959], Braithwaite [1953], Reichenbach [1938], Feigl [1970] and most logical empiricists since the time of the Vienna Circle.

2.2. *Inventionist View*

An articulate minority of philosophers of science has opposed the received view, and has insisted that the study of the context of invention is indeed a legitimate concern of philosophy, and that a logic of invention can (in principle) be explicated or reconstructed. The most vociferous of these *inventionists*, in the last few decades, has been Hanson [1961; 1965; 1967a; 1967b] who has proposed (*inter alia*) a "retroductivist" schema of invention, following Peirce [1931]. More recent sympathisers, in their various ways, with this viewpoint include Achinstein [1970], Post [1971], Simon [1977], Lakatos [1976] – mainly in regard to invention in mathematics – and Holton [1978]. However, despite a largely un-acknowledged liberalising of outlook lately, the *anti*-inventionists still have the numbers on their side.

The reason why inventionism has not been taken more seriously within philosophy of science is that none of the inventionists has succeeded in giving a clear conceptual analysis of the points at issue – especially the distinctions and connections between logical and empirical questions – so as to demonstrate the legitimacy, and the rationale, of the enterprise. I aim to repair this defect.

2.3. *Inconoclast View*

A third, more iconoclastic stance is to deny the reality of the invention/appraisal dichotomy altogether. Kuhn ([1964], p. 9) has cast doubt on the distinction, at least in its traditional form; and Feyerabend ([1975], p. 167) "disposes" of it entirely. He tries to achieve this by emphasising the *continuities* between the inventing and the appraising procedures employed by a scientist (" ... we are dealing with a single uniform domain of procedures all of which are equally important for the growth of science." ibid.). That is, certain historical, psychological or political influences may affect not only the genesis but also the acceptance of a particular scientific theory.

I have some sympathy with this view of Feyerabend's, for it makes clear that an empirical reconstruction of the context of appraisal is a legitimate

and important enterprise.[4] Conversely, I would insist that a *logical* reconstruction of the context of invention is licit and valuable; indeed, this is my
central thesis. It may well be the case that, as Feyerabend says, the same
empirical factors can play (causal) roles in both the invention and the appraisal of a theory. Similarly, I shall urge that the same logical structures
(what I shall call below *advancement/enhancement* arguments) may function
in both contexts. However, I can still admit certain pragmatic or temporal
differences between inventing and appraising. Further, I want to insist on
a clear distinction between logical and empirical questions. It is *this* distinction which some critics (Feigl [1970], Hempel [1965a]) have accused Kuhn
and Feyerabend of overlooking; insofar as they do ignore it, I disagree with
them. They have been accused also of ignoring or confounding a related
distinction, namely that between description and prescription − e.g., in their
respective tendencies to equate historical (empirical, descriptive) and normative (logical, prescriptive) accounts of science.[5]

Now, *contra* Kuhn and Feyerabend, I *accept* both the invention/ap-
praisal distinction and the empirical/logical distinction. But *contra* Popper,
Reichenbach *et al.*, I *deny* that the former distinction can be based upon
the latter.

3. SALMON: TWO THESES

Let me introduce my approach to these questions by referring to the writings
of Reichenbach's distinguished pupil, Wesley Salmon − who was also my
teacher.

3.1. *Psychology and Logic*

In his [1970c], Salmon gives a very clear account of Reichenbach's original
distinction between the contexts of discovery and justification. In particular,
he insists that the two contexts may have items in common: it is a frequent
misrepresentation of Reichenbach's view to maintain that the contexts
exclude each other. In Salmon's words:

The context of discovery consists of a number of items related to one another by psychological relevance, while the context of justification contains a number of items
related to one another by (inductive and deductive) logical relevance. There is no reason
at all why one and the same item cannot be both psychologically and logically relevant
to some given hypothesis ([1970c], p. 72).

Characteristically clear as this is, I think Salmon's formulation may be misleading nonetheless. If by the 'context of discovery' (or 'invention') is meant the class of items relevant to the "hitting upon" of a hypothesis, then it seems evident enough that this class may include *both* empirical items (e.g., events) and logical items (e.g., propositions).[6] That is, if we are interested in giving a *causal* account of the inventing of a hypothesis H, we may provide a description in, say, psychological terms of what events "led up to" the scientist's inventing or "hitting upon" H; but this in no way prevents us from giving, as well, a *logical* account of the matter, in terms of certain *arguments*, having H as a conclusion, which the scientist may have contemplated or reflected-upon in the course of his/her inventing-activity. (This assumes that the scientist is rational, and has the goal of a *plausible H*.) The same applies to the context of appraisal. In short, it is not *only* the case – as Salmon rightly notes – that the same item (e.g., an event or a proposition) may belong to *both* invention and appraisal; but *further*, it is the case that both empirical and logical items may belong to the *same context*. The distinctions between invention and appraisal, on the one hand, and between empirical and logical items, on the other hand, are not parallel – as presumed by the advocates of the received view – but rather are orthogonal conceptually to each other.

This insight – that the logical/empirical distinction can be drawn with equal facility in *both* the context of invention and the context of appraisal – suffices to rebut the charge of psychologism – i.e., of conflating empirical and logical matters – which Popper, for one, has levelled against the inventionist program (Popper [1959], p. 31). The point is that in *either* context one can raise: (a) *empirical* (e.g., psychological) questions about the *causes* of a scientist's actions, including his/her inventings-of, as well as acceptings-of, rejectings-of, or perseverings-with, certain hypotheses; and (b) *logical* questions about the inferential relations between a hypothesis and guiding considerations, plausibility considerations and data statements. (I shall define these various "considerations" shortly.)

3.2. *Two-Stage Bayesian Analysis of Appraisal*

This leads me to the second, extremely important, contribution made by Salmon in his [1970c] and elsewhere:[7] his interpretation of *Bayes' Theorem* in his rational reconstruction of the context of appraisal. According to Salmon's construal, *two* stages of this context are to be distinguished.

There is the *initial* stage, when the hypothesis H has been invented, but not yet tested. At this stage, the scientist (call him/her 'S') has to decide whether H is worth proceeding-with, or whether it should simply be discarded without more ado. This decision will be based upon judgements of utilities, which in turn must take into account the probability-value of H at this prior-to-test stage. This *prior probability* of H is an estimate of its (initial) *plausibility*. The value assigned to this probability is to be determined according to the interpretation of probability adopted by S: on a frequentist view of probability, which Salmon favours, various plausibility considerations may be invoked, such as deductive relations with other hypotheses whose probability has already been assessed in the light of evidence. In addition, and most significantly for my present account, considerations of simplicity, symmetry and analogy with other already-successful hypotheses are highly relevant to the estimate of prior probability for H. Of course, other authors as well as Salmon have emphasised the great importance of such plausibility considerations in the appraisal of a hypothesis; but he has provided a formalisation of their role by means of his Bayesian schema, which I shall sketch below. The plausibility considerations enter into what I call *enhancement arguments* for H: typically, these are inductive arguments with the plausibility considerations among the premises, and with H as the conclusion. I shall exemplify them shortly. I call this prior-to-test phase the *enhancement* stage of the context of appraisal. (Some authors, such as Laudan | [1980], refer to this as "the stage of pursuit". I avoid this term for several reasons, but especially because once the context of appraisal has been entered, H is no longer being pursued; it has already been captured; and S is trying to decide whether it is worth adopting and putting to work, or whether it should be tossed back into the jungle of wild conjectures.)

The *second* stage of the context of appraisal is that of *confirmation* of H. If the scientist S has decided, in the light of his/her estimate of utilities (which in turn involve the prior probability of H), to attempt the confirmation of H, he/she usually will set up an appropriate test, in which H is "confronted with nature" in some way — e.g., in the form of an experiment or set of observations using an H-D schema. The key question before the scientist at this stage is: what is the appropriate probability-value to assign to H after the test-result is known? Salmon answers: this probability-value, called the *posterior* (to test) probability of H, may be construed as a function of several other probabilities, including the (previously assessed) *prior* probability of H.

The relationships among the values of the posterior probability of H, the prior probability of H, and two other probabilities, namely the probability of the observed outcome E, given H (called the *likelihood* of H) and the probability of E independently of H's truth-value, are summarised, according to Salmon, in the following schema derived from Bayes' Theorem:[8]

$$\text{posterior probability} = \frac{\begin{array}{c}\text{prior probability}\\\text{of } H\end{array} \times \begin{array}{c}\text{probability of } E,\\\text{given } H\end{array}}{\begin{array}{c}\text{probability of } E, \text{ independently}\\\text{of } H\text{'s truth-value}\end{array}}$$

In an *H-D* situation, where *H* *entails* a particular test-outcome, the value of the *likelihood* expression in the numerator of the right-hand side will be either unity or zero, depending on whether the outcome predicted by *H* turns out to be observed or not. In case the test-result is negative, yielding a zero likelihood for *H*, the whole R.H.S. will become zero, and thence the posterior probability of *H* will be zero – i.e., *H* will have been falsified. The Bayesian schema thus neatly exhibits the *modus tollens* falsification procedure central to Popperian deductivism. More interesting for my purposes is the case where the predicted test-outcome is observed, i.e., where the evidence supports or confirms *H*. In such instances, the value of the likelihood factor becomes 1, so that the posterior probability of *H* is given by the ratio of the prior probability of *H* to the probability of the observed outcome *E*, independently of *H*'s truth-value.

It can be seen that according to this Bayesian schema, which is proposed by Salmon as a realistic improvement upon the *H-D* schema, the prior probability of *H* enters in a crucial way into the assessment of its (non-zero) posterior probability, that is, the probability-value assigned to *H* after a favourable test-outcome. This posterior probability can then, of course, become a prior probability for *H* in the *next* set of tests; and so on. It is evident that the original prior probability of *H* continues to play a role in the assessment of later values of its posterior probability. This prior probability results from enhancement arguments for *H*, as I have noted. Thus enhancement arguments function in a very important role in the context of appraisal for *H*, if one adopts Salmon's Bayesian schema for appraisal.

4. ADVANCEMENT, ENHANCEMENT, CONFIRMATION

4.1. *"Plausibility Considerations" and Advancement/Enhancement Arguments*

Having set the stage by introducing the notion of enhancement arguments for *H* in the context of appraisal, I now want to propose the idea of *advancement arguments* for *H* in the context of invention. Typically, these will be inductive arguments employing as premisses considerations of simplicity, symmetry and analogy with other already-successful hypotheses (or *models*), among other propositions – i.e., premisses of the same type as Salmon's "plausibility considerations" – and having as conclusion the hypothesis invented. I shall give examples of such arguments below (Section 6).

Central to this explication of the logic of invention is the thesis that advancement arguments for an H in the context of invention may *also serve* as enhancement arguments for the same H in the context of appraisal. This possibility can be demonstrated by examples (Section 6). The *rationale* for the logical explication of invention is then immediately apparent, for it follows directly from this that the hypotheses advanced by advancement arguments will be *plausible* hypotheses – their plausibility being guaranteed by the very arguments that advanced them, and that, in the context of appraisal, now serve to enhance them. Once this insight is clearly achieved, the initial appeal of the orthodox anti-inventionist thesis, with its insistence on the categorial – rather than simply temporal-pragmatic – distinction between the contexts of invention and appraisal, vanishes; as does the attraction of the thesis that epistemology is properly concerned solely with the rational reconstruction of appraisal.

4.2. *Rationale of Inventionism*

Let me develop briefly this point about the rationale of my enterprise. The scientist S wants to, and sometimes does, hit upon an initially-plausible H. My account explicates how he/she does this rationally, when he/she rationally does it. In short, my account "makes sense of" invention (or aims to), just as the H-D and other accounts "make sense of" appraisal (or aim to), by attempting to explicate the *logical bases* of the scientist's rational actions – appraisings, inventings. The overlap between invention and appraisal – i.e., the identity of certain advancement and enhancement arguments – expresses the insight that "making sense of" or rationally reconstructing the two contexts often involves very much the same sorts of analyses.

Linked with this is the recognition that the study of invention may *contribute* to the reconstruction of appraisal, through the common focus of both contexts upon *plausible H*'s. That is, the question (a) "What rational ponderings led S to judge H plausible?" and (b) "What rational ponderings led S to invent H?" often may have the *same* answer, namely: "S's ponderings of a particular advancement/enhancement argument". Hence, the study of such arguments in the context of invention can contribute to the rational reconstruction of the context of appraisal, whose *rationale* consequently will apply to the logical explication of invention.

This refutes a common objection to inventionism, which has been expressed by Laudan [1980] among others. He acknowledges that, in what he calls "the older programme" of scientific methodology, induction by enumeration played the *dual* role of advancing and enhancing (confirming, in that programme) hypotheses. But he cannot see a comparable role for a logic of invention in "the newer programme", which is characterised by *H-D* and related methodologies, and in which higher-order hypotheses of great scope, less directly connected with evidence than in the "older" situation, are to be appraised. My reply to Laudan and to others of this ilk is to point out that in this "newer programme" [if, indeed, the distinction between older and newer methodologies can be drawn so nicely], *induction by analogy*, and other varieties of inductive reasoning, play the *dual* role of advancing and enhancing (making plausible, in this programme) hypotheses. Contrary to Laudan's view, the current situation is not all that different in principle from his "older" one. The rationale of a logic of invention which he admits applied then, also applies now. Note especially that, now as then, the logic of invention, like the logic of appraisal, remains *inductive*. This point is central to my thesis. To deny it, as Simon [1977], e.g., does, is indeed to expose oneself to Laudan's criticisms, as I show below (Section 8.3). In affirming the crucial role of induction in invention, I reject the view that invention is non-inductive (Simon), as well as the contention that inventionism lacks a rationale (Laudan).

The rationale of my enterprise can be summarised thus: the aim of a logic of invention is to explicate the advancement of *plausible* hypotheses. The search for such a logic is worthwhile if the search for plausible hypotheses is worthwhile. This latter search is worthwhile because plausible hypotheses are what science wants (or what human beings, for whatever motives, want of science), and what a logic of appraisal is concerned to appraise.

4.3. *Logical and Empirical Conditions for Reconstructions*

We have seen that the context of appraisal embraces two distinct stages, namely *enhancement* (prior to test) and *confirmation* (posterior to test). Before appraisal there is invention, comprising the stage of *advancement* of the hypothesis. Each of these three stages – confirmation, enhancement, advancement – will admit of *both* logical and empirical reconstruction, as I emphasised above (Section 3.1). Now I want to make explicit the typical or normic conditions for such reconstructions. Let the hypothesis H be the object of attention of a scientist S, and assume that S is rational, honest and motivated to attain a well-confirmed hypothesis. Two background philosophical theses are also assumed in this reconstruction, namely inductivism and rational determinism.

The pair of normic conditions for the two reconstructions of the stage of *confirmation* of H are:

> *Logical Condition I*: Certain propositions O_i, available to S after test, are logically related to H in such a way that they confirm H.

> *Empirical Condition I*: Certain events E_j are associated with the propositions O_i, and cause S to tentatively accept H.

Note that the events E_j typically will be quite different from the events or states-of-affairs described by the O_i. Thus the E_j might include those events which comprise S's *pondering* of the O_i (e.g., these E_j might be certain of S's brain-events); whereas the O_i normally would be statements of evidence or other considerations supportive of H.

Preceding the confirmation stage of the context of appraisal, there is the stage of *enhancement* of H, as I have remarked. The pair of normic conditions for the two reconstructions of this enhancement stage are:

> *Logical Condition II*: Certain propositions P_i, available to S before test, are logically related to H in such a way that they enhance H.

> *Empirical Condition II*: Certain events F_j are associated with the propositions P_i, and cause S to judge H plausible.

As before, the F_j will include S's pondering of the P_i (plausibility considerations for H).

This explication now proceeds into the context of invention of H. In the context of invention, closely connected with the pair of conditions in the enhancement stage of appraisal, is a pair of normic conditions for the two reconstructions, logical and empirical, of the advancement of H, namely:

> *Logical Condition III*: Certain propositions Q_i, available to S before the invention of H, are logically related to H in such a way that they advance H.

Empirical Condition III: Certain events G_j are associated with the propositions Q_i, and cause S to invent H.

Once more, the events G_j will include S's pondering of the Q_i — which I call "guiding considerations" for H. The study of these events G_j and of their causal connections with the inventing of H is a matter for appropriate empirical disciplines like psychology and the history of science — as the received view rightly has it. The same remark applies to the empirical study of the events F_j and E_j, and their causal connections with S's respective judgings that H is plausible and tentatively acceptable.

Quite apart from these empirical inquiries, however, it seems surely clear that the study of the propositions Q_i and their logical relations with H is just as properly philosophical as is the study of the plausibility considerations P_i and their logical relations with H. This latter study is acknowledged, by Salmon and others, as an important component of the logical explication of the context of appraisal. It seems *prima facie* capricious then to deny that the study of the logical relations between Q_i and H is an important constituent of the logical explication of the context of invention. This conclusion is reinforced by the recognition that in many cases the P_i and the Q_i are identical, as are the advancement and enhancement arguments for H which employ them as premises.

The distinctions and connections among the logical and the empirical components of the reconstructions of advancement, enhancement and confirmation are summarised in the following table, where the symbol $\longrightarrow$ indicates an inductive (logical) relation, and the symbol $\Longrightarrow$ indicates a causal (empirical) relation.

CONTEXTS

		INVENTION	APPRAISAL	
		Advancement	Enhancement	Confirmation
COMPONENTS	LOGICAL (propositions)	$Q_i \longrightarrow H$	$P_i \longrightarrow H$	$O_i \longrightarrow H$
	EMPIRICAL (events)	$G_j \Longrightarrow S$ invents H	$F_j \Longrightarrow S$ judges H plausible	$E_j \Longrightarrow S$ tentatively accepts H

Here is a biographic scenario for H. A scientist S, in search of a hypothesis H, contemplates a set of propositions Q_i, and this contemplative activity causes him/her to hit upon (invent) H. Having invented H, S next inquires whether it can be enhanced: how plausible is it? In judging this, S takes account of whatever plausibility considerations P_i for H are available. It would not be unusual for these P_i to include, if not coextend with, the guiding considerations Q_i. If S's estimate of the prior probability of H, based on P_i, yields a sufficient utility-value to warrant testing, he/she conducts a test. If H is not falsified by the test, its posterior probability is calculated by S, using a Bayesian schema, in which the numerator of the right-hand side of the equation is the prior probability value for H based on the P_i. The relations between the Q_i and H, and the P_i and H, can be logically reconstructed respectively as the advancement and the enhancement arguments for H.

As this stage of my analysis, let me attend briefly to the difficulties in Hanson's attempted rational reconstruction of invention.

5. HANSON: THE THREE-FOLD FAILURE OF RETRODUCTION

Following an idea of Peirce [1931], Hanson attempted to explicate the logic of invention ("discovery") by means of a concept of "retroduction" or "abduction". Peirce viewed this form of reasoning as distinct from both deduction and induction, although he was never very clear about its distinguishing features. The examples of "retroduction" given by Hanson seem to me to be straightforward cases of inductive reasoning – in particular, induction by *analogy* (as against "straight" induction by enumeration).

5.1. *Legitimacy*

The first of three weaknesses in Hanson's treatment of invention was his failure to give an adequate clarification of the conceptually orthogonal or independent relationship between the logical/empirical and the appraisal/invention dichotomies. This failure meant that the *legitimacy* of his program of rationally reconstructing the context of invention was always in question; it was thus understandable that his proposals were taken less seriously than they deserved by the majority of philosophers of science.

5.2. *Dual Status of Some Considerations*

Second, although he did recognise the role of advancement arguments in the context of invention, Hanson was troubled by the apparent *dual status* of some considerations, such as particular symmetries and analogies, which seemed to function in *both* invention and appraisal. As he acknowledged, such considerations " ... are reasons *both* for proposing that H will be of a certain type and for accepting H." ([1961], p. 27). He apparently was looking for *logically different* procedures in invention and in appraisal; his unfulfilled hope seemed to be that some special sort of reasoning – retroduction – would turn out to operate specifically in the context of invention, whereas another kind of reasoning – induction, for instance, as it functions in the *H-D* method – would characterise the context of appraisal. (In contrast, I insist that the "logic of invention" is mainly induction.)

Hanson's difficulty seems to result, in part at least, from his lack of anything like Salmon's *two-step* Bayesian account of the logic of appraisal, with its enhancement stage as well as its confirmation stage. This lack on Hanson's part enabled his critics, including Salmon ([1967a], p. 113), to charge him with offering only *plausibility* considerations (enhancement arguments) for the hypothesis H, rather than, as he purported, genuine invention (advancement) arguments for H. I have urged above that, far from the dual role of some considerations in the advancement and enhancement of an H being an embarrassment for an inventionist program, this is exactly what one would expect from an adequate logical reconstruction of the two contexts.

5.3. *Reconstruction of Appraisal, Not of Invention*

The objection that Hanson was in fact reconstructing appraisal arguments, not the logic of invention, was strongly reinforced by the third defect in his account, namely his presentation of a "retroductive schema" for the invention of a hypothesis H:

(1)   Some surprising, astonishing phenomena p_1, p_2, p_3 ... are encountered.

(2)   But p_1, p_2, p_3 ... would not be surprising were a hypothesis of H's type to obtain. They would follow as a matter of course from something like H and would be explained by it.

(3)   Therefore there is good reason for elaborating a hypothesis of the type H: for proposing it as a possible hypothesis from whose assumption p_1, p_2, p_3 ... might be explained.

Hanson [1961], p. 33.

On examination, this "retroductive schema" turns out not to be a reconstruction of an invention argument at all. It sheds no light whatever on what
considerations suggested or pointed to (advanced) H or H-type hypotheses;
rather, in Premiss (2) H is taken as somehow *given*, and the retroduction
schema merely illustrates a method of deciding whether it is *plausible*. That
is, retroductive reasoning may play a part in the context of appraisal *after H*
has been hit upon; retroduction does not function in invention.[9]

Even to call the above Premiss (2) a *plausibility* consideration for H may
be granting too much. All Premiss (2) really does is to specify a *necessary
condition* for H's being a candidate-explanation of the phenomena p_i. Something stronger than mere satisfaction of this necessary condition would
seem required before H could be given the title 'plausible'. This "something
stronger" has been provided, in fact, by Achinstein [1970] in his elaborated
retroduction schema, which he calls the "explanatory mode of inference".
This elaboration, which takes account of background information, and of
H's explanatory power in comparison with that of rival hypotheses, does
provide a basis on which H may be deemed plausible (or the "most plausible"
of the available hypotheses); but it in no way meets the problem just noted.
The new H still appears *explicitly*, and without any account of its genesis,
in a premiss; so that Achinstein's proposal does not succeed at all in transforming Hanson's reconstruction into an *invention*-schema.

6. ADVANCEMENT ARGUMENTS

What is needed is a schema for invention which shows how the hypothesis
follows inferentially from some premisses, in which it does not already appear
explicitly. The premisses themselves must admit of plausibility estimates,
at least to the extent that their sources or bases are clearly identified. A
valuable clue for the construction of such a schema has been given by Hanson
[1961], despite his confusions, in his mention of the role of *analogy* in
invention. I now want to propose that an important class of (but not all)
advancement arguments are essentially arguments by analogy, these being
seen as a significant class of (correct) *inductive* arguments. In such arguments,
the key premiss is a statement of an analogy or similarity between the phenomenon of interest (the problem situation, which sometimes may have the
status of an *anomaly*) and some other phenomenon which has the role of
a *model*. From this and other premisses, and utilising a rule of analogy
(Section 7.2, below), the hypothesis H is inferred inductively. (There may
also be deductive stages in the overall inference-chain.) I shall give three

examples of such advancement arguments, and also one of a non-analogical advancement argument relying on simplicity principles, plus a lemma using a mixture of both. Before I do so, in order to minimise misunderstanding, let me try to make clear the roles of these examples.

6.1. *Roles of Examples*

An example of something which is claimed not to exist demonstrates, at the very least, the falsity of the existence denial. Popper has stated unequivocally: "There is no such thing as a logical method of having new ideas, or a logical reconstruction of this process." ([1959], p. 32). This claim is very strong and therefore highly falsifiable. All that is needed to falsify it is *one* acceptable case of a logical reconstruction of the invention of a new idea or hypothesis. In fact, I believe that reconstructions of this kind are possible in a great many cases of invention of hypotheses (plausible hypotheses, at any rate) — which is just to say that I believe that many inventings are rational. However, I do not make the bold, falsifiable and very probably false claim that all cases of invention are logically reconstructible. (For that matter, not all *appraisings* are rational, or logically reconstructible.)

I offer four examples of logical reconstructions of hypothesis-inventions, plus a lemma, none of which seems to be particularly contrived or far-fetched. The *point* of the offer also is fourfold. (a) That such reconstructions are possible will certainly suffice to falsify Popper's (and others') *general* denial of this possibility. (b) The reasonableness and straightforwardness of the examples should serve as plausibility considerations for my thesis that such explications are readily constructible in a great many cases of hypothesis-invention. (c) It should also be apparent on inspection that any of the advancement arguments illustrated could serve equally well as an *enhancement* argument for the H in question.[10] (d) Finally, by analysis of the examples, I aim to provide a catalogue of the leading principles and guiding considerations which function in the construction of advancement arguments.

6.2. *Philosophy and History*

Note that I do not claim that, as a matter of historical fact (!), any particular scientist did follow the advancement argument sequences illustrated here. My present thesis is philosophical rather than historical; and, as I have urged, the rationale for a philosophical reconstruction of invention derives directly from the epistemological rationale for the study of appraisal. A philosopher may say that, in some historical case, a hypothesis H was accepted by a scientist, or scientists, on "good" or on "bad" (e.g., irrational) grounds. This in effect is to state a definition of "good grounds for accepting H" – that is, to proffer a definition of "scientific rationality" or "scientific morality", as Lakatos [1978b] would put it. I regard this as proper work for philosophers of science; it exhibits the *normative* aspect of our enterprise. Similarly, I remark that there may be "good" and "bad" (e.g., irrational) grounds for *inventing H*. Such "good" grounds or reasons in fact correspond with, and may be the same as (or derive from, via enhancement arguments) the "good" reasons for accepting H. A definition of scientific rationality, in the form of a logical reconstruction of the context of appraisal (which would include explication of the guiding considerations Q_i – cf. Section 7.2), is *ipso facto* a definition of good reasons for inventing H. The *epistemological* value of explicating the logic of appraisal/invention is not affected by the circumstance that, historically, some hypotheses may have been accepted for other-than-good reasons, just as some hypotheses may have been invented irrationally. Indeed, such historical judgements themselves must rest on philosophically prior reconstructions, as Salmon [1970c] points out.

6.3. *Examples of Advancement Arguments*

Example I: The Rutherford Atom [11]

(Q_1)   Atoms are small entities whose structure is unknown, but they appear to be both dense and diffuse, since some (alpha) particles collide violently with them, while other (beta) particles pass uneventfully through them.

(Datum)

(Q_2)   Atoms in this respect *are analogous to* the Solar System, through which most foreign bodies (e.g., extra-system comets) pass uneventfully, although some collide with the Sun.

(Analogy claim)

(Q_3)   The Solar System has the structure of a (relatively) small, massive nucleus (Sun) orbited by much lighter bodies (planets).

(Datum)

*

* I use the convention of a double line between premises and conclusion to indicate inductive inference, and a single line to represent deductive inference.

(H)   Atoms are structurally similar to the Solar System, consisting of massive nuclei of relatively small diameter, orbited by much lighter particles.

Example II: The Inverse-Square Gravitation Law for Planets [12]

(Q_1)   A planet orbiting the Sun is constrained in its path by a radial force whose quantitative relation to the length of the radius vector is unknown.

(*Leading principle*)

(Q_2)   In this respect, a planet orbiting the Sun *is analogous to* a weight whirling on the end of an anchored string, in which string a centripetal force constrains the weight in its path.

(*Analogy claim*)

(Q_3)   The force F in an anchored string attached to a whirling weight is given by

$$F \propto r/T^2$$

where r is the length of the string and T is the period of the rotation.

(*Huygens' Principle*)

(H_1)   The (gravitational) force constraining a planet in its orbit around the Sun is proportional to r/T^2, where r is the planet's radius vector and T is its year.

(Q_4)   The year of any planet is related to its radius vector by the formula:
$$T^2 \propto r^3$$

(*Kepler's Third Law of Planetary Motion*)

(H_2)   The (gravitational) force constraining a planet in its orbit around the Sun is proportional to r/r^3; that is, to
$$1/r^2$$

Example IIIa: Structure of DNA Molecule — Analogical Advancement
Argument [13]

(Q_1) The structure of the DNA molecule is unknown, but one of its
major chemical constituents is a form of nucleic acid.

(Background information)

(Q_2) In chemical composition, DNA *is analogous to* TMV (tobacco
mosaic virus), which also has a form of nucleic acid as a major
chemical constituent.

(Analogy claim)

(Q_3) The TMV molecule is helical in structure.

(Datum)

───

(H) The DNA molecule is helical in structure.

Example IIIb: Structure of DNA Molecule — Simplicity-based Advancement
Argument [14]

(Q_1) The structure of the DNA molecule is unknown, but it is probably
crystalline.

(Early X-ray diffraction data from Wilkins)

(Q_2) Crystals have regular (symmetric) structures

(Background information)

(Q_3) "The simplest form for any regular polymeric molecule ... [is]
... a helix."

(Simplicity claim — Watson ([1968], p. 131))

───

(H_1) The structure of the DNA molecule is a helix.

(Q_4) The thickness of the DNA molecule indicates that it is composed
of more than one chain, i.e., that it is a compound helix.

(X-ray data)

(Q_5) The simplest compound helix is a double helix.

(Simplicity claim)

(H_2) The structure of the DNA molecule is a double helix.

Note 1: It is highly compatible with my thesis that Watson made telling use of simplicity considerations and related aesthetic principles as *plausibility considerations* for the double-helix hypothesis, *after* its invention. Thus he remarks: "A structure this pretty just had to exist" ([1968], p. 161).

Note 2: Watson also made use of an advancement argument, utilising both analogy and simplicity principles, to derive the hypothesis that the TMV molecule is helical. This hypothesis was quickly confirmed, and then served as Premiss Q_3 in Example IIIa above. An admissible derivation would be the following:

Lemma: The TMV Molecule is Helical in Structure [15]

(Q_1) The structure of the TMV molecule is unknown, but it is observed to grow rapidly.

(Datum)

(Q_2) In this respect, TMV *is analogous to* crystalline substances, which also exhibit growth.

(Analogy claim)

(Q_3) Crystals grow by a mechanism of accretion.

(Background information)

(H_1) TMV also is a crystal, which grows by a mechanism of accretion.

(Q_4) The growth-rate of TMV is much higher than that of a simple crystal.

(Background information)

(H_2) TMV is a complex crystal which grows through possession of "cosy corners" in which material rapidly accretes.

(cf. Watson ([1968], p. 94))

(Q_5)   " ... the simplest way to generate cosy corners ... [is] ... to
            have the sub-units helically arranged."

                                                    (Watson, *ibid.*) (*Simplicity claim*)

(H_3)   The TMV molecule is helical in structure.

7. LOGICAL RECONSTRUCTION OF INVENTION

7.1. *Paradigmatic Structures of Advancement Arguments*

Comparison of Examples IIIa and IIIb suggests that there are at least two
main types of advancement arguments, viz. (a) *analogical* ones, employing a
rule of analogy and having an analogy claim among the premisses; and (b)
arguments using a rule of *simplicity*, and having a simplicity claim among the
premisses. Types (a) and (b) are perfectly compatible with each other, as their
mixture in the Lemma illustrates.

(a) The typical or paradigmatic structure for an *analogical* advancement
argument is this:

(Q_1)   Identifies or locates the problem phenomenon/situation, often
            by highlighting some significant or striking feature of it (atoms
            are dense and diffuse; DNA is probably crystalline; etc.).

(Q_2)   Statement of a particular analogy between the striking feature of
            the problem phenomenon and some property of a model.

(Q_3)   Description of some other interesting feature of the model.

(H)   Ascription of the latter feature to the problem phenomenon
            also.

Often the argument is elaborated, as in Example II, by addition of premisses and inferences to further H's.

(b) *Simplicity-based* advancement arguments, like Example IIIb, include premisses which specify some set of constraints on admissible H's; then a simplicity claim, together with a rule of simplicity, serves to select one from among the available admissible H's.

I proceed to list the chief inference rules and leading premisses of advancement arguments.

7.2. *Principles of Advancement Arguments*

The insight underlying the rational reconstruction of invention is that sequences of ideas G_j (cf. Section 4.3 above) are not blind, but rather correspond to logical arguments. These advancement arguments, as we have seen, typically conform to certain inference rules, and employ as premisses certain "leading principles"[16] or general claims about the world, as well as more specific empirical statements, including singular analogy and simplicity claims. (A similar idea is expressed by Holton ([1978], p. 9), in his notion of *themata*.) The most important of these rules and principles are the following.

(a) *Inference rules.* These rules are to be understood as *inductive* inference principles. As with other inductive principles (and, for that matter, deductive principles like non-contradiction) I cannot validate them. However, I think a strategy of *vindication* is the correct approach to the problem of induction.[17] Since inductive inference, as expressed in various inductive rules, e.g., analogy and simplicity (*q.v.*), plays an important role in the context of appraisal — not only in confirmation, but, more significantly for my purposes, in enhancement — it follows that the logic of invention is in no worse case than is the former, by virtue of its reliance on induction. (See also Section 8.3 below.)

(i) Rule of *Simplicity*: "Choose the simplest of the available admissible hypotheses." Application of this rule depends, of course, on the particular definition of simplicity invoked in a given case. Although some cases are fairly straightforward, others are far from obvious. See helpful discussion of this by Hesse ([1974], pp. 223–257) and Sober [1975]. Important particular simplicity principles are listed under "leading principles" below.

(ii)   Rule of *Analogy*: "From a similarity (analogy) between two phenomena in a significant respect, infer a similarity between them in some other respect pertinent to the problem phenomenon." This might be viewed as a sub-species of an over-arching simplicity principle which, in effect, would assert that in nature heterogeneity is minimised, or similarities are maximised. See Hesse [1966], especially on this.

(*b*) *Premisses*. These include certain "leading principles", primarily versions of simplicity, together with more specific claims, such as statements of particular analogies. On an empiricist epistemology, all these *guiding considerations* (Q_i) are seen as knowable *a posteriori* and falsifiable *in principle*; although some may be very resistant to refutation, as the history of science shows.

(i)   *Leading* (*simplicity*) *principles*:
- *Isotropy* with respect to space: "There are no privileged directions in space". This is in contrast, for instance, to the Aristotelian doctrine that every body has its "natural place", toward which it strives.
- *Homogeneity* with respect to space: "There are no privileged locations in space" (such as "natural places" or absolute origins). These and other simplicity considerations (e.g., symmetry) were highly relevant to Einstein's Special Theory of Relativity (cf. esp. Hesse (op. cit.)).
- *Empirical Symmetry*: expressions of this might be "Forces in nature are equally balanced"; "Shapes tend to mirror each other"; "Systems strive toward homeostasis". Newton's 3rd Law of Mechanics, "To every action there is an equal and opposite reaction" expresses the symmetry principle; and its role in the advancement/enhancement arguments for the double helix structure of the DNA molecule has been illustrated. (cf. Shubnikov and Koptsik [1974]).
- *Parsimony*: "Entities are not multiplied beyond necessity" (Ockham's Razor). This is a principle of conservatism which pervades science, and serves to curb temptations to excess, such as proliferations of *ad hoc* hypotheses.

From time to time, one or another of these principles, in some particular application, may turn out to be false and be abandoned; perhaps to reappear in some more sophisticated guise. Again, principles may clash with each other: thus postulating a new charged particle to restore "balance of charges" in accord with the symmetry principle may seem to constitute a "multiplication beyond necessity" of entities. In such cases, one or other of the clashing principles will be abandoned. I call these clashes "conceptual anomalies"; their role in scientific change is crucial, but not widely understood.

(ii) *Particular analogies*. In each of my examples of analogical advancement arguments (Section 6.3), the second premiss Q_2 was a statement of a particular analogy between the phenomenon of interest (e.g., atoms, DNA) and some other phenomenon (e.g., Solar System, TMV) about which certain things were already well-known. In such cases the latter phenomenon is said to have the role of a *model* for the former. (Hesse [1966]).

The systematic search by the scientist S for suitable premisses Q_2, stating particular analogies which will serve in advancement arguments for H, may be likened to a process of *scanning* (e.g., by radar) in search of appropriate models. To develop this metaphor, the scanning radar beam symbolises a constraining requirement of analogical relevance which hits upon, and echoes back from, suitable candidate models or analogues of the problem phenomenon − as a radar beam of a certain wavelength and amplitude will detect certain targets and ignore anything else. Obviously the models have to be already in the field of the scanner − that is, within S's body of concepts. This sets a straightforward empirical limitation on what S can rationally invent; but note that his/her field of concepts need not be limited to consciousness, and that complex concepts can be *constructed* out of simple ones, consciously or unconsciously.[18] Once a suitable model for the problem phenomenon has been recognised, it forms the basis for a particular analogy premiss, and from this, together with other premisses (including guiding considerations), and in accord with various formal rules (including analogy), S is enabled to set up an advancement argument for H.

This *scanning* procedure is a metaphor for an "algorithm" that is automatic only in a highly-qualified sense. It is consistent with the inductive character of most advancement arguments, and certainly provides no support for a rationalist, aprioristic view of invention.[19]

8. IMPROVEMENTS

8.1. Sources of the Q_i: "Scientific World Views"

At this point it may be objected that my proposed rational reconstruction of invention is all very well, but that it sheds no light on the genesis of the premisses Q_i in advancement arguments. Why does S pick or hit upon *this* striking feature of the problem phenomenon, or *that* particular analogy, in preference to the enormous variety of other candidates available? Have I really advanced (sic!) the problem beyond Hanson's account, in which H itself appears "out of the blue" in a premiss of his retroduction schema?

Now note that this central question is not confined to the invention of hypotheses. Under-determination of conclusions by premisses is a problem in the context of appraisal also. The "appraisal-counterpart" of the above question might be expressed: why does S hit upon some particular plausibility considerations (P_i) for H? Additionally, in appraisal, there is the question why S chooses particular deductive consequences of H in setting up his/her tests, and ignores the remaining members of the infinite set of propositions entailed by H.

The answers to all these questions are primarily philosophical, and secondarily empirical (psychological/historical). In the first place, they hinge upon each individual's (e.g., scientist's) possession of what may be called roughly a "Scientific World View" (SWV). For each person, this SWV consists of a vast set of beliefs about the world, many of them unconscious, un-articulated or implicit in more overt ones, which serve to define what seems to him/her "natural" as well as "odd" in his/her experience, and thence to pick out or highlight striking or significant features of problem situations, as well as of possible models therefor.[20]

In the second place, *how* people, including scientists, acquire their SWVs is an empirical question. I suggest that they are learned *a posteriori*, as a result of a huge array of experiences, including formal education and exposure to whatever may be the prevailing background SWV(s) of a particular culture at a particular time. These "background SWVs" themselves may have been confirmed by evidence to some extent, but may express also the influence of various other factors and beliefs, including religious, metaphysical, aesthetic, economic and political ones. As well as the set of beliefs comprising his/her SWV, each person has also a set of cognitive abilities, aptitudes, capacities and skills, including, e.g., "analogy recognition ability", which may contribute to the on-going formation of the SWV.

The empirical answer to the question, "Why did S pick out *this* particular feature of the problem situation as *the* striking or significant one?", would be along these lines: S's encounter with the problem caused him/her to reflect upon it, and initial reflection, in the light of S's SWV and cognitive abilities (e.g., "analogy recognition ability") caused S to hit upon and then ponder detailed propositions Q_i. These Q_i in turn logically advanced a certain H, and S's pondering of them caused him/her to invent that H — as I have already outlined. In summary, the philosophical genesis of the Q_i is accounted for in terms of the scientist's SWV; and the empirical fact of his/her possession of this particular SWV, together with a particular set of cognitive abilities, causally explains his/her tendency to "zero-in" on certain features of problem situations, as well as providing an empirical back-stop to a regress of "why" questions.

Answers along similar lines, philosophical and empirical, can be elaborated for the same sorts of questions in the context of appraisal, such as: 'why did S pick out these particular plausibility considerations for H?'; and, 'why did S select these particular observational consequences of H for test-purposes?'. (The answer to this last question may well include reference to all manner of *pragmatic* factors, such as availability of instruments, accessibility of certain kinds of data, and the like.)

8.2. *Novelty*

Any genuine logical reconstruction of invention must be compatible with the advancement of *novel* hypotheses, and inductive advancement arguments of the kind outlined here are fully capable of this. Principles such as analogy and simplicity which function in these arguments enable the conclusion to "go beyond" the content of the premises. In analogical advancement arguments, this ampliative process is one of *induction by analogy* (in comparison to induction by enumeration), and typically amounts to applying in a new context or to a new phenomenon a property already familiar in an old context. Strikingly consistent with this is the view that a degree of *openness of texture* (Waismann [1951]) or of *partial definition* of theoretical concepts by reduction sentences or correspondence rules (Carnap [1963–7]; Braithwaite [1953]; Feigl [1970]; *et al.*) is a *desideratum* in science, although ironically the authors mentioned have mainly held the orthodox, anti-inventionist view.

8.3. *Invention and Induction*

Of course, it is true that even an automatic deductive algorithm can advance novel hypotheses in one sense of 'novel', namely, ones that have not been thought of before. But this is of little interest to science, since what matters is the advancement, not of *any* new hypothesis, but of *plausible* new hypotheses. *Inductive* advancement arguments, of the kind illustrated above, are of value in science precisely because they give to the invented hypothesis an initial degree of inductive support, or plausibility, which accompanies the hypothesis as it enters the context of appraisal. They are thus in sharp contrast to deductive algorithms, such as Simon's [1977] "BMA" and "HSA" (see below), which apparently provide no plausibility value at all for the hypotheses ("patterns") which they advance. Since Simon's purportedly deductive approach to invention is directly opposed to my inductive reconstruction, let me discuss it briefly.

Simon's BMA ("British Museum Algorithm" – named after the monkeys who were alleged to have used it to reproduce the volumes in the British Museum) is a "blind" data-scanning procedure, which uses a systematic trial-and-error strategy of pattern-search. The HSA ("Heuristic Search Algorithm") is more sophisticated, employing such concepts as "same" and "successor" to vastly reduce the number of patterns eligible for trial. These are examples of what Simon misleadingly calls "law discovery processes" (ibid., p. 331) – misleadingly, since a *law*, as ordinarily understood, is a generalisation; whereas these algorithms are merely recoding devices. A law is more than a summary of data; so a genuine law discovery process must be an *ampliative* process. He then defines a *normative theory of scientific discovery* as "a set of criteria for evaluating law-discovery processes" (ibid.).

Now, a *normative* theory or set of criteria is, by definition, *goal*-oriented, as Simon himself insists (ibid., p. 328). The goal of a law discovery process, in science at any rate, is surely the discovery of hypotheses which will turn out to be highly-confirmed in the context of appraisal; that is, *plausible* hypotheses. It follows that a normative theory of scientific discovery will be a set of criteria for evaluating the efficacy of various law discovery processes with respect to the goal of discovering plausible hypotheses. Some patterns yielded by such processes will be better than others, according to criteria specified by the normative theory – that is, some patterns will be more parsimonious, elegant, symmetric and so forth than others, since these are characteristics which, in the context of appraisal, render hypotheses more plausible. Indeed, it is difficult to imagine what *other* sorts of criteria might be realistically specified by Simon's "normative theory of scientific discovery".

Simon has sought to divorce "law discovery" from the appraisal of the law (hypothesis) invented thereby; and thence to claim that this "discovery" process is non-inductive. I contend that such a divorce makes his "law discovery" quite pointless and idle scientifically. He acknowledges that there must be norms or criteria for evaluating law discovery processes, yet this very acknowledgement seems to imply that scientific desiderata, such as the features of a hypothesis that make it plausible, must be invoked to give point to the law-discovery enterprise. (My remark here is not unlike Laudan's [1980] criticism of *some* current approaches to the logic of invention — cf. Section 4.2, above). Against Simon, I maintain that law-discovery (hypothesis-invention, in my preferred terminology) actually involves ampliative, non-demonstrative inference in accord with chosen rules and leading principles, expressing certain scientific desiderata (e.g., guiding/plausibility considerations), and is thus inescapably inductive in character. Were it not, scientific discovery would indeed be as irrational as Popper *et al.* have sought to persuade us. To his credit, Simon explicitly disavows this Popperian line; to be consistent, he should then espouse mine.

9. CONCLUDING REMARK

Not merely one, but two kinds of irrationalism pervade current views of scientific invention. According to the received orthodoxy of Popper, Reichenbach, Braithwaite and like-minded philosophers, inventive and other creative activities "transcend" rationality, and hence cannot be rationally reconstructed. This false intuition is then somehow tied in with a charge of "psychologism", directed against any attempts to give a logical explication of invention. The latter charge would be more telling, did it not rest on conflations of invention with psychology, and of appraisal with logic. In fact, as I have shown, the invention/appraisal distinction is quite independent of the psychology/logic dichotomy. The context of invention embraces *both* logical and psychological (as well as other empirical) questions; as does the context of appraisal.

The other irrationalist approach to invention is that of the iconoclasts — Kuhn, Feyerabend and their backers. These authors deny any interesting distinctions between invention and appraisal, logical and empirical questions, description and prescription. The vehemence of their opposition to older orthodoxies perhaps diminishes the impact of their many good insights — that hypothesis-appraisal has an empirical component, for instance, and that among the empirical influences which may bear upon the accepting or rejecting of a hypothesis are factors which have little connection with rationality. While acknowledging this, I still want to insist that rational deliberations are among the major causal influences on hypothesis appraisings, as on their inventings; and, less daring than the iconoclasts, I try to stay a little longer with the traditional distinctions, such as that between logic and psychology, which they scorn.

The moral of my tale, unfashionable as it may be in such circles (and perhaps in an earlier, Viennese one, for different reasons), is that scientific change, including invention, is largely rational. Whatever other empirical factors causally influence movements in science, the *rational reflections* of scientists upon certain problems, empirical and conceptual, are among them. To ignore these rational reflections is arbitrarily and confusedly to truncate science and the philosophy that studies it. To attend to them requires explicating their *logical bases*, a task which embraces and gives point to the logical reconstruction of the context of invention.

Macquarie University

NOTES

[1] A shorter version of this paper was presented at the annual conference of the Australasian Association for the History and Philosophy of Science, University of Melbourne, August, 1979. For valuable criticism and stimulation of my present ideas on invention, as they evolved over a number of years, I am grateful to Wesley Salmon, Ben Rogers, Hugh Mellor, Mary Hesse and Heinz Post.

[2] "The way, for instance, in which a mathematician publishes a new demonstration, or a physicist his logical reasoning in the foundation of a new theory, would almost correspond to our concept of rational reconstruction; and the well-known difference between the thinker's way of finding his theorem and his way of presenting it before a public may illustrate the difference in question. I shall introduce the terms *context of discovery* and *context of justification* to mark this distinction. Then we have to say that epistemology is only occupied in constructing the context of justification" (Reichenbach [1938], pp. 6–7).

[3] Throughout this paper, I assume – *contra* Popper and other deductivists – that there is a non-empty class of *correct inductive inferences*. I follow Salmon ([1967a], p. 26) in urging that Popper's "corroboration" involves induction, and that inductive inference is inescapable in the context of appraisal. For this reason, and further, because (as I show) appraisal and invention intersect, the logic of invention is no worse off than the logic of appraisal by virtue of its reliance on induction.

4 With Lakatos [1978b] I would add the proviso that this must depend on a *logical* reconstruction of the methodology of appraisal.

5 See, for example, Musgrave [1982].

6 I assume that Salmon, in the passage quoted, was offering exemplifications, not definitions, of the two contexts. Of course, if 'discovery' is *defined* as psychological, and 'justification' is *defined* as logical, then my position would be simply that *this* discovery/justification distinction is conceptually orthogonal to *my* invention/appraisal distinction.

7 Notably his [1967a]; and also his [1968a].

8 Salmon [1967a], p. 117, [1970c], p. 79.

9 The fact that, in retroduction, the H turns up *explicitly* in a premiss, so that its advent remains quite unexplained by the schema, has been noted by several authors – e.g., Saunders [1972], Frankfurt [1958].

10 This is not to deny the possibility that some advancement arguments do not, in fact, enhance their H's – a point made by my friend Professor Ben Rogers.

11 Reconstructed from an account in Hoffman [1947].

12 Reconstructed from an account in Hanson [1961].

13 Reconstructed from anecdotes in Watson [1968].

14 Ditto.

15 Ditto – cf. especially pp. 93–94.

16 Some of these have been called "metaphysical principles". As an empiricist, I avoid this title, because it seems to bestow on such propositions a spurious immunity to falsification. Insofar as they are claims about the world, these principles are indeed falsifiable, although often they are abandoned with great reluctance – cf. Kepler's battle to save the principle of uniform circular celestial motion. Insofar as they are unfalsifiable, their status is rather that of inference rules expressing determinations to reason in certain ways. The two are not, by the way, epistemically independent; but nor are they simply interchangeable.

17 On the general notion of vindication, see Feigl [1950]. On the vindication of induction, see Reichenbach ([1938], pp. 348–357); Salmon ([1961a] and [1967a], pp. 52–54); Clendinnen [1966] and especially his "Rational Expectation and Simplicity", this volume.

18 This notion of *construction* of complex ideas out of simple(r) ones is not too dissimilar in spirit from Hume's well-known account (see, e.g., his *Treatise*, Book I, Part I). The conjecture that a scientist has in his/her unconscious a vast array of concepts, hypotheses and potential problem-solutions, only some of which ever become conscious, has been suggested, along lines highly compatible with my own outline here, by Paul Meehl in his remarkable essay, 'Psychological Determinism and Human Rationality: A Psychologist's Reactions to Professor Karl Popper's "Of Clouds and Clocks"' (Meehl [1970]).

19 The finding of a suitable model, for use in the analogy-premiss of an advancement argument (Q_2 in my examples), by the scientist S is automatic *only if* two *contingent* conditions on S are fulfilled. These are that (a) there *is* a suitable model in S's field of concepts; and (b) S is good enough at model-recognition to find it during his/her lifetime. This innocuous sense of 'automatic' carries no suggestion of the guarantees associated with a *rationalist* "invention-algorithm".

20 My notion of a "Scientific World View" has close affinities with Kuhn's [1962]

"paradigms", Lakatos' [1973a] "scientific research programmes" or Holton's [1978] set of "themata". In order to avoid unnecessary methodological commitments associated with some of these terms, and because my SWV is intended to be rather wider and looser than any of these other sets of propositions, I have breached the spirit of Occam's Razor by introducing a further term for the same general idea.

JOHN SAUNDERS AND JOHN NORTON

EINSTEIN, LIGHT SIGNALS AND THE ε-DECISION

1. INTRODUCTION

It is now three-quarters of a century since the publication of Albert Einstein's 'On the Electrodynamics of Moving Bodies'. Early in the spring of 1905, when the idea of his work on relativity was "still a mere concept", Einstein anticipated which part of his work would capture the imagination of his contemporaries. In a letter of March 6th to his friend Conrad Habicht he wrote of "the electrodynamics of moving bodies by the use of a modification of the theory of space and time" and that "the purely kinematic part of this work will undoubtedly interest you".[1] At the heart of the kinematic part of Einstein's work lies his discourse on the concept of simultaneity.

The aim of this essay is to explore certain aspects of the rationale and methodology of Einstein's revolutionary treatment of the "definition of simultaneity" of events at two distant points within one inertial reference frame. As will become clear, we relate these matters to those aspects of the philosophic writings of Wesley Salmon which are concerned with the nature of conventions in physical theories.

The immediate value of Salmon's contribution to the debate over the philosophic thesis of the conventionality of simultaneity is that he has drawn attention to the logical interrelations between the concept of simultaneity and a number of physical hypotheses concerning light signals, the determination of a one-way velocity of light, the limiting character of the velocity of light, and the constancy of the velocity of light.[2] To achieve our aim in this paper, we attempt to use these logical insights heuristically to focus directly on a number of historical theses which have often been misunderstood in contemporary appraisals of conventionality in Einstein's 1905 account. We argue that the view of conventionality held by Salmon is necessary to an understanding of the methodology of Einstein's discourse but that it is not sufficient for a complete appreciation of the decision made by Einstein to adopt a unique definition of distant simultaneity. We also maintain that an understanding of some of the concepts that are central to Reichenbach's and Salmon's writings on the nature of simultaneity in a relativistic universe is essential if one is not to misconstrue the prominent status which Einstein

101

Robert McLaughlin (ed.), What? Where? When? Why?, 101–127.
Copyright © 1982 *by D. Reidel Publishing Company.*

and many others have given to light in special relativity. In particular, we have in mind here the concept of the first signal, the concept that space will not allow signals to propagate faster than a certain maximum speed.

In agreement with Hans Reichenbach and others, Salmon seems to suppose that it is unproblematic that Einstein believed that, in some important sense, the relation of distant simultaneity is conventional.[3] After all, it is often pointed out, the title of Section 1 of the Kinematical Part of Einstein's paper is 'Definition of Simultaneity' and, as a first step in his reformulation of classical kinematics, Einstein himself emphasises the role of a "definition" in his brief account of distant simultaneity.

If at the point A of space there is a clock, an observer at A can determine the time values of events in the immediate proximity of A by finding the positions of the hands which are simultaneous with these events. If there is at point B of space another clock in all respects resembling the one at A, it is possible for an observer at B to determine the time values of events in the immediate neighbourhood of B. But it is not possible without further assumption to compare, in respect of time, an event at A with an event at B. We have so far defined only an "A time" and a "B time", but no common "time" for A and B. The latter can now be defined in establishing by definition that the "time" required by light to travel from A to B equals the "time" it requires to travel from B to A.[4]

This rule for determining distant simultaneity, known as standard signal synchrony, seems quite innocent, yet it is pregnant with problems of great significance. Even on the basis of a straightforward reading of this passage, how should we understand Einstein's assertion here that distant simultaneity depends crucially on a "definition"? Is this "definition" equivalent to what is now termed "conventionality"? Is it open to negotiation? Could Einstein have chosen other non-standard definitions for distant simultaneity without this leading to contradictions? Or was he forced, in 1905, to the definitions that he used by sound physical considerations?

Further, what can we say about Einstein's decision to use light signals as the basis of his definition? Can we conclude from this that light signals have special properties possessed by no other signals and that the theory which Einstein produced has as much to say about the properties of light as it does about those of space and time? Or is it the case that other signals could equally well have been used by Einstein in his definition? And, if this is so, has the importance of light in the theory, and indeed in the universe as a whole, been exaggerated as the result of Einstein's decision to feature light signals so prominently?

2. NOMOTHETIC CONVENTIONS AND THE CHOICE OF ϵ

In order to prepare the ground for the consideration of these questions, we will begin by giving a clear account of some of the philosophical issues raised by the standard version of the thesis of the conventionality of distant simultaneity. The purpose of this is twofold. First, it will help to clarify exactly what is now claimed in the thesis. Second, it will enable us later to delineate the nature of Einstein's seminal contribution to the discussion, to give some insights into the reasons behind Einstein's choice of a particular "definition" of simultaneity and to determine the nature of the role that light signals play in that definition.

Let us again consider the two points at A and B, between which we transmit light signals in empty space. On a space-time diagram (two-dimensional), points A and B are represented by two indefinitely extended world-lines perpendicular to the axis of space and light signals as straight diagonal lines intersecting them (Figure 1). Let event E_1 be the emission of a light signal at (A time)t_1 from A, event E_2 the reflection of the light signal at B and event E_3 its return to A at (A time)t_3. In accord with Einstein's "definition", an event E at A that is simultaneous with E_2 at B is one that occurs at (A time)t_2, where $t_2 = t_1 + \frac{1}{2}(t_3 - t_1)$.

To arrive at this determination of t_2, we have made use of what is looked upon in the thesis of the conventionality of simultaneity as an arbitrary stipulation. That is, we have made use of Einstein's decision to equate the time required for light to travel from A to B to the time it requires to travel from B to A. If the distance AB is taken to be equal to BA, a moment's reflection shows that this decision is equivalent to the assumption that the speed of light from A to B is equal to the speed of light from B to A.

That such a stipulation is necessary is the essence of the conventionalist thesis. It is needed to break a logical circle that we enter when we try to judge the simultaneity of distant events. To see this we note that in Einstein's approach, the determination of the simultaneity of events at two distant points A and B is equivalent to the synchronisation of a clock at A with one at B. Now, if it were possible a priori to synchronise the A clock with the B clock then it would be a simple matter to determine experimentally the speed of light as it travels from A to B, the so-called "one-way speed of light". We transmit a light signal from A to B and note the "time" of emission at A and the "time" of arrival at B. We then divide the distance between A and B by the transmit "time" (the difference in the two values) and arrive at the one-way speed of light between A and B. Conversely, if it were possible to

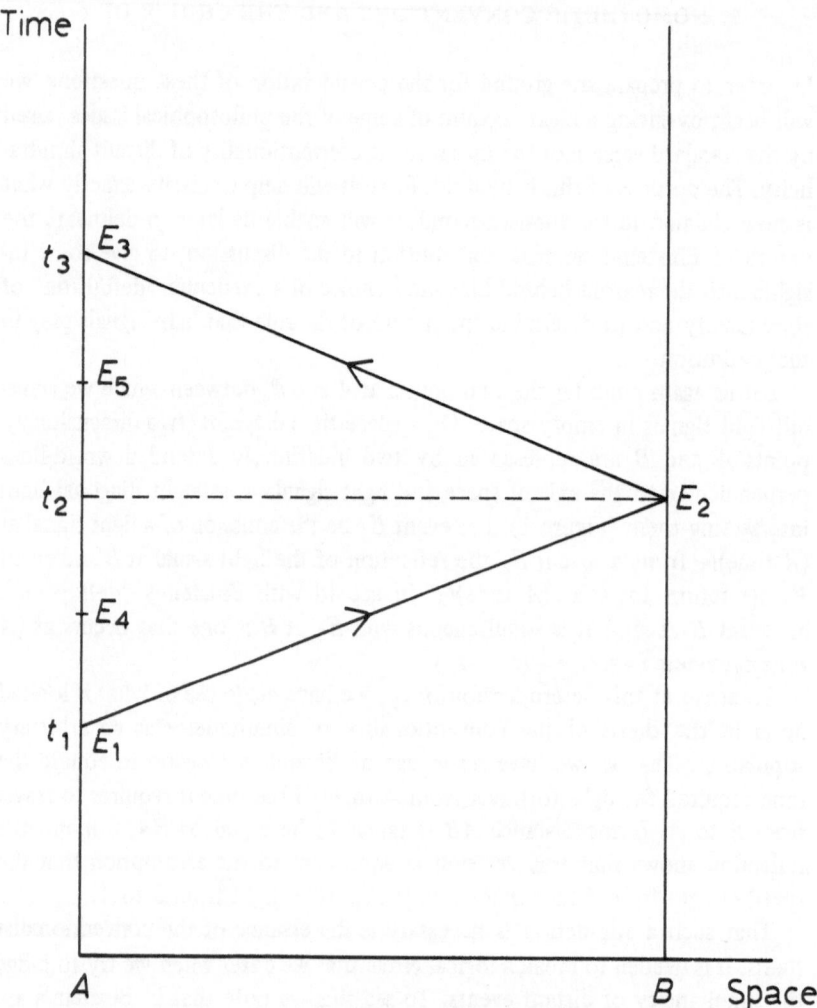

Fig. 1. Light signal synchrony.

determine a one-way speed of light between A and B, we could reverse the procedure and so synchronise the clocks at A and B *a posteriori*. As Reichenbach noted,

We are faced with a circular argument. To determine the simultaneity of distant events we need to know a velocity, and to measure a velocity we require knowledge of simultaneity of distant events. The occurrence of this circularity proves that simultaneity is not a matter of knowledge, but of a coordinative *definition*, since the logical circle shows that a knowledge of simultaneity is impossible in principle.[5]

In Einstein's approach, we break the circle by stipulating that the one-way speed of light is equal to the so-called "two-way speed of light", the average speed at which light travels from A to B and back to A. This two-way speed of light can be determined experimentally without becoming involved with the circular argument. Its determination requires only knowledge of times t_1 and t_3, which can be read off the one clock at A, and the distance AB. But this is not the only way to break the circle. According to the conventionalist thesis, it is possible, without contradiction, to stipulate that *any* event between E_1 and E_3 at A is simultaneous with E_2 at B. In a compact notation introduced by Reichenbach, denoting the A-time of the event at A simultaneous with E_2 at B by t_2, we can write:

$$t_2 = t_1 + \epsilon(t_3 - t_1)$$

and a direct expression of the conventionalist thesis is that we can arbitrarily choose the value of ϵ in the interval

$$0 < \epsilon < 1$$

without the possibility of contradiction.[6]

The choice of $\epsilon = \frac{1}{2}$, however, is traditional and is known as "standard signal synchrony". The choice of ϵ other than $\frac{1}{2}$, in the interval $(0, 1)$, leads to anisotropy in the one-way speed of light, even though the round trip speed for light, the two-way speed of light, always remains constant with the familiar value of c. It is a somewhat surprising result that such a "non-standard" choice of ϵ leads to no physical contradictions. Winnie has shown that it is possible to rederive the basic kinematics of special relativity, in particular the Lorentz transformation, with the value of ϵ undetermined in the interval $(0, 1)$ without leading to physical contradictions.[7] This suggests strongly that no experiment could ever lead to the need to select one particular value of ϵ.

Salmon has drawn attention to the fact that we need not employ light for the signalling process used to determine the timing of distant events. Any other signal, with a two-way speed that is isotropic and constant over space and time, would serve equally.[8] Repeating the above arguments for such a signal, we come to the conclusion that the knowledge of any one-way velocity is impossible, unless we are prepared to make some arbitrary stipulation that would enable us to synchronise distant clocks.

So far we have not considered other methods of synchronising distant clocks. Could we not synchronise distant clocks, as suggested by Ellis and Bowman,[9] by the use of a clock, transported slowly from A to B – in fact, transported infinitely slowly in the limit in order to eliminate the effects of relativistic time dilation? According to the conventionalist thesis, any such method must contain, at least implicitly, a stipulation equivalent to the adoption of standard signal synchrony. In the case of slow clock transport, this is taken to be the stipulation that clocks infinitely slowly transported away from a point remain in standard signal synchrony with clocks at that point. If a non-standard signal synchrony rule has been adopted, the readings of such clocks will systematically deviate with distance from standard signal synchrony.[10]

Salmon has put considerable effort into elucidating the unusual status of the convention involved in the conventionalist thesis.[11] On the one hand, he recognises the existence of "trivial semantic conventionality". At its simplest, this comprises an agreement between people that certain terms have certain meanings, for example, that the term "simultaneity" means "sameness of time" or that a centimetre be the hundredth part of a metre. Although essential if human discourse is to occur, such conventions are of little interest for the problems of space and time raised here.

Salmon's claim is that the conventionality involved in the question of distant simultaneity is nontrivial because it rests on – indeed it is only made possible by – the truth of certain non-trivial hypotheses. Salmon singles out here the hypothesis that there is an upper limit to the speed of signal propagation. We can look upon the conventionality thesis itself as entailed by a physical law: it is impossible to determine the one-way speed of light experimentally.

This is, of course, not the only possible statement of the thesis. We can, however, readily see that it is equivalent to the conventionalist thesis. If we knew as a matter of nonconventional fact that two distant events were simultaneous, we could use them to synchronise two distant clocks, and, in turn, use these to determine the one-way speed of a light signal propagated between them. If it is given that we cannot determine the one-way speed of light, then we can conclude by *modus tollens* that we can never know as a matter of nonconventional fact that two distant events are simultaneous.

This recasting of the conventionalist thesis is of great interest. The initial treatment of the thesis was based on the discussion of the existence of an unavoidable uncertainty in our knowledge of the time coordinate of distant events. Now the thesis has been rewritten as a law-like statement of new and interesting knowledge about space and time and the propagation of signals. Considerations such as these lead us to suggest that such conventions be called Nomothetic Conventions.

It would appear that in this new formulation, the conventionalist thesis is now more readily open to confirmation or refutation. With a view to defending the thesis from refutation, in a recent essay Salmon examines nine different proposed experiments which seek to determine a one-way speed of light.[12] He proceeds to argue that each fails because it contains an implicit convention, at a crucial juncture, which is equivalent to the covert choice of standard synchrony. This strongly suggests that there is no experiment which could ever yield a determination of the one-way speed of light. In a sense, the paper lays down the challenge of finding such an experiment but, in the end, suggests that it would be possible to find some nontrivial convention equivalent to a synchrony rule, implicit in any that may be put forward. Indeed, if we are willing to allow the range of conventions Salmon claims to be equivalent to the adoption of standard synchrony, it is hard to imagine any such convention-free experiment.

We may wish to question to what extent this state of affairs is satisfactory. Salmon seems to be asking us to accept the thesis on the grounds that no one can refute it, rather than presenting us with an indisputable proof of the thesis. In questioning this strategy we should recognise the peculiar nature of the statement of the thesis concerned. It is an example of what Sir Edmund Whittaker called "postulates of impotence" which assert "the impossibility of achieving something, even though there are an infinite number of ways of trying to achieve it".[13] Indeed, this formulation is not very different from that of the foundation postulate of the special theory of relativity in the form: it is impossible to determine an absolute velocity in empty space. Again, one formulation of the second law of thermodynamics states that it is impossible to create a perpetual motion machine. When this is used as a statement of the second law, we cannot, of course, use other formulations of the second law to verify that no such machine is possible. Rather we have to resort to plausibility arguments in which a convincing number of candidates for perpetual motion machines are examined and shown to be lacking on one ground or another. We can conclude that Salmon's tactic is only one possibility in the circumstances, but that he is in good company in using it.

We have already seen that Salmon singles out the hypothesis that there is a maximum speed at which signals can propagate in space as being of special importance to the conventionalist thesis. We shall now examine the relationships among this hypothesis, the conventionalist thesis and special relativity, since this hypothesis is going to play a key role in a later section of this paper. The value of this maximum speed is, of course, usually taken to be the speed of light. This has led Reichenbach to call light a "first signal".[14]

We claim that the hypothesis that there is a maximum speed for signal propagation forms an essential part of the conventionalist thesis. Indeed, it is logically entailed by it. But we claim it is a weaker thesis, for it does not in turn entail the conventionalist thesis.

That this hypothesis is a necessary part of the conventionalist thesis follows readily from our earlier discussion. If, as in the Newtonian world, there were no upper limit to signal speeds, then we would, in theory, have little trouble in judging the timing of distant events. We could unambiguously and, of course, nonconventionally determine what event at a distant point B was simultaneous with a given event at a point A by sending out a signal which propagates at infinite speed from the event at A to B. In the Newtonian world, such a signalling operation could be carried out by monitoring the variations in the gravitational field at B resulting from a change in position of the field-producing masses at A.

An important point to note here is that the actual value of this maximum speed is nowhere near as significant as the fact that such a maximum speed exists. If, for example, the speed of a first signal was greater than that of light then, referring to Figure 1, we should find that a first signal emitted from E_4 would still arrive at E_2 and then return to A at E_5. The effect of this would be to reduce the range of choice of events at A that can be conventionally chosen to be simultaneous with E_2 at B (to the interval separating E_4 and E_5) but not to eliminate this conventionality. Similarly, it would still be impossible to determine the one-way speed of light nonconventionally.

We should note, of course, that if the speed of a first signal were to differ from that of light, we would encounter serious difficulties in the wider realms of special relativity, in particular, in its treatment of electromagnetism, as we shall see later.

We now go on to show that the hypothesis of a maximum speed for signal propagation does not entail the conventionalist thesis. This can be seen by considering a hypothetical universe in which, by some means, judgements of distant simultaneity can be made nonconventionally. Clearly, none of the formulations of the conventionality thesis discussed above could obtain in such a universe. In particular, we could use our nonconventionally determined knowledge of distant simultaneity to determine a one-way speed of light.

However, it would be possible in such a universe for there to be a maximum speed for signal propagation, without logical contradiction. An example of such a universe is the one described by special relativity, with the added proviso that, due to knowledge from some other source, events seen to be simultaneous using clocks synchronised by standard signal synchrony are, as a matter of fact, simultaneous.

The status of the hypothesis of the maximum speed of signals within special relativity is not completely clear in the literature. Grünbaum argues that, at least historically, it was an assumption of the original theory and it should not be looked upon as a result of the velocity addition law.[15] Salmon writes of the hypothesis as a consequence of the theory,[16] presumably because of his demonstration that the admission of faster-than-light signals into special relativity leads to paradoxical closed causal loops.[17] Elsewhere he notes that the result is well established experimentally, in so far as it has been found impossible to accelerate particles beyond the speed of light.[18]

Setting aside the historical question raised by Grünbaum, we look upon the hypothesis as of fundamental importance to our understanding of the special theory of relativity. Whether one regards it as a postulate or a consequence of the theory, is ultimately a matter of taste. It is possible within a multiply connected logical structure, such as special relativity, to interchange axioms and theorems in such a way that what, in one version, is a hard-won result appears as a postulate in another.

We shall claim, however, in a later section, that as far as the heuristics of special relativity are concerned, there are significant gains to be made if we look upon this hypothesis as a postulate of the theory. In fact it will be seen to replace the light postulate as one of the two fundamental postulates of the theory.

To conclude this account of the conventionalist thesis, we shall now turn to discuss what we see as a weakness of the thesis, one that is of particular relevance to an understanding of the historical emergence of relativity and Einstein's account of simultaneity.

We maintain that the conventionalist thesis is bound by a narrow epistemology. It is clear that in straightforward empirical terms there is an uncertainty in our judgements of distant simultaneity, an uncertainty that can only be resolved by an empirically arbitrary stipulation. It is also true that, in terms of physical predictions, the infinitely many different formulations of special relativity outlined by Winnie, are equivalent.[19]

The problem with this epistemology manifests itself in the failure of the conventionalists to allow that there can be factors involved in the choice of theories that are non-empirical in character, factors that appear purely on a theoretical level. In what follows, we discuss two aspects of the range of ε-formulations of special relativity in order to give examples of the types of non-empirical factors that may be active in the question of theory choice. First, the formulation of a non-standard version of special relativity is significantly more complex and cumbersome than the standard formulation. Second, on a theoretical level, the thesis postulates a state of affairs that is very hard to come to terms with in the framework of the heurstics of special relativity: that is, in nonstandard formulations, the one-way speed of light is anisotropic.

In the view of Reichenbach, and apparently Salmon, these aspects of the thesis in no way predispose us to select the standard synchrony formulation as being, in any factual way, a better representation of reality. The preference of relativists for the standard synchrony version of special relativity is to be seen solely as a result of the fact that it is more convenient to work with, because its equations and concepts are simpler.

This is clearly admissible within a naive empiricist epistemology, but runs seriously against the heuristics of relativity and our experience of what have been fruitful lines of attack in physics. It is a commonplace expectation amongst physicists that the laws of nature seem to find simple mathematical expression. If a proposed law can only find inelegant mathematical expression, then the physicist will have some grounds to doubt it. Certainly, he will not rest easy until he can find an elegant mathematical formulation for it. Indeed, mathematics has come to be seen as a kind of "natural language" for the laws of nature and we expect that all such laws, if they are true, will find simple expression in this language. This type of thinking has been elevated to a central position in the theory of relativity through the principle of general covariance. This principle requires that all true laws of nature must be capable of being written in a generally covariant form. It makes no requirement on the physical content of the law but only on how it is written, that is, its mathematical expression.

The suggestion that the one-way speed of light is anisotropic in the non-standard formulations of special relativity is one that, on a realist interpretation of the theory, is inconsistent with the heuristics of relativity. What justification can there be, we might ask, for setting up preferred directions in space for light propagation. The conventionalists, of course, would reply that such anisotropy is not claimed to occur in any physical sense; it is a purely theoretical construction. Such a reply is, we believe, unsatisfactory. It is a dangerous tactic to dismiss as unimportant a reliance on theoretical entities whose properties run fundamentally counter to the heuristics of the overall theory. This is especially so in the case of the theory of relativity, a theory whose fundamental principles are concerned as much with the empirical content of its laws as with their theoretical formulation.

These objections are not directed at the conventionalist thesis itself but at the equivalence of the ε-formulations of special relativity that it suggests. This leads us to ask whether there might be another formulation of special relativity, which incorporates the conventionalist thesis, but one whose theoretical structure is not open to the objections which we have made here. We speculate that such a formulation may result from an attempt to write special relativity with two-way quantities only, which would avoid using an undetermined ε in the formalism. Such a reformulation may well not look at all like traditional formulations, because a large number of the fundamental quantities of such formulations, including many velocities and lengths, are one-way quantities. If such a reformulation of the theory could be carried through successfully, it might well have important ramifications for the conventionalist thesis and special relativity itself.

None of these objections to the ε-formulations of special relativity which result from the conventionalist thesis actually refute the thesis. Rather, they serve as warnings against being too complacent about the apparent success of the verificationist arguments that underlie the thesis and they must be borne in mind when we ask such historical questions as why Einstein chose standard signal synchrony and not another.[20]

We shall argue, in the next section, that, whatever the merits of the conventionalist assessment of standard signal synchrony, it does not accurately reflect the rationale for Einstein's historical decision to choose $\epsilon = \frac{1}{2}$ in 1905.

3. THE HISTORICAL CHOICE OF $\epsilon = \frac{1}{2}$

In contrast to the elegance of many philosophical narratives concerning the conventionality thesis, any understanding of Einstein's approach to the simultaneity of distant events must draw upon a rich matrix of historical and methodological concerns. For example, by means of his thorough reading and discussion of Henri Poincaré's *Science and Hypothesis*, with his friends Maurice Salovine and Conrad Habicht in their self-styled "Academy" in Berlin, Einstein became aware of Poincaré's earlier arguments for the need for "a new rule for the investigation of simultaneity". In a paper entitled 'La Mèsure du Temps' published in 1898 and again, more briefly, in *Science and Hypothesis*, Poincaré argues that we cannot legitimately speak of the simultaneity of two events occurring in two different places without introducing some "convention".[21] In part support of the more general conclusion that "absolute space, absolute time and even geometry are not conditions which are imposed on mechanics", Poincaré claims that,

1. there is no absolute space and we only conceive of relative motion; and yet in most cases mechanical facts are enunciated as if there is an absolute space to which they can be referred;
2. there is no absolute time; when we say two periods are equal, the statement has no meaning and can only acquire meaning by a convention;
3. not only have we no direct intuition of the equality of two periods, but we have not even direct intuition of the simultaneity of two events occurring in two different places.[22]

Poincaré suggests a convention which provides a criterion of simultaneity and is based on the "natural" procedure of signalling between the two places by means of light rays. He claims that this "rule" could be based on the "postulate that light has a constant speed and in particular that its speed is the same in all directions". This view of the criterion of distant simultaneity as a convention is consistent with Poincaré's broader thesis of the generally conventional nature of our choice of the concepts and procedures of physics. It seems to us that Einstein's 1905 discourse on distant simultaneity incorporates important aspects of Poincaré's account. Of particular relevance here is that, first, Einstein, like Poincaré, considers only signal synchrony. Second, as we have seen, Einstein's "definition" that the "time" required for light to travel from A to B equals the "time" it requires to travel from B to A, is equivalent to the assumption that the speed of light from A to B is equal to the speed of light from B to A, and this, in turn, is equivalent to Poincaré's "postulate" that the speed of light is the same in all directions. Einstein, like Poincaré, adopts the device of using light signals and the properties of light to map out and even define the properties of space and time.

In an important way, however, Poincaré's conventionalist account does not encompass Einstein's sense of "definition". Pertinent here is the *epistemology* of Einstein's general approach to the electrodynamics of moving bodies. This was formulated in accord with a pervasively empiricist and critical methodology drawn together from the traditions of David Hume, the empiricist, and Ernst Mach, the positivist.[23] It is our contention that it is this eclectic methodology which leads Einstein to decide that, in the strict absence of any *facts* which guarantee that light travels with the same speed in different directions in empty space, he would need a "further assumption" that this was definitely so if he wished to make use of light signal synchrony.

This sense of Einstein's "definition" is, we think, in accord with the primary notion of "convention" in the contemporary thesis of conventionality according to which the choice of $\epsilon = \frac{1}{2}$ is not seen as a matter of *fact*.

In view of the historical uncertainty about the isotropic propagation of light in empty space, why does it seem so "natural" to Poincaré, and presumably Einstein, to establish distant simultaneity by means of light signals? In his 1921 Princeton Lectures on *The Meaning of Relativity*, in answer to the criticism that relativity theory gives "without justification, a central role to the propagation of light, in that it founds the concept of time on the law of propagation of light", Einstein replies that

in order to give physical significance to the concept of time, processes of some kind are required which enable relations to be established between different places. It is immaterial what kind of processes one chooses for such a definition of time. It is advantageous, however, for the theory, to choose only those processes concerning which we know something certain. This holds for the propagation of light *in vacuo* in a higher degree than for any other process which could be considered, thanks to the investigations of Maxwell and H. A. Lorentz.[24]

In our view, this retrospective account accurately pin-points the dominant paradigmatic setting in which Einstein's ideas on these matters were thought through. An understanding of just what could be known as a matter of fact about the propagation of light is clearly crucial to an understanding of Einstein's account of simultaneity. As is well known, Einstein's special theory of relativity is based on two conjectures which, in the 1905 account, are given the systematic status of postulates. The second of these Einstein later called the "L-principle".[25] His initial 1905 formulation of it reads as follows, "light is always propagated in empty space with a definite velocity c which is independent of the state of motion of the emitting body".[26] He does not say here what he understands by "empty space" but later it becomes clear, in his derivation of the transformation equations of space and time coordinates, that it is homogeneous and isotropic. If this is the case, we may take it that a corollary of the L-principle is that the speed of light is the same in all directions. Einstein's second statement of the L-principle in his 1905 paper enriches this picture in an unexpected way. "Any ray of light moves in the 'stationary' system of coordinates with the determined velocity c, whether the ray be emitted by a stationary or a moving body", he next writes.[27]

What is this "stationary" system? It is "a system of coordinates in which the equations of Newtonian mechanics hold good". That is, it is an inertial system which is apparently named only "in order to render the presentation more precise and to distinguish this system of coordinates verbally from others which will be introduced hereafter".[28] However, Einstein bases his development of the electrodynamics of moving bodies on Maxwell's theory for "stationary" bodies and initially supposes that the fundamental equations for the structure of the electromagnetic field given by this theory hold in the "stationary" system. These same equations had earlier been taken by Maxwell and others late in the nineteenth century as being true only for a system of reference in an ether that is homogeneous and quiescent. On this construal we can equally well take it as a corollary of the L-principle that the speed of light is the same in all directions.

It seems reasonably clear from this that the L-principle *defines* the very property of the propagation of light which is crucial to Einstein's approach to distant simultaneity. The general grounds for Einstein's initial confidence in, and postulation of, the L-principle are therefore of relevance to our enquiry. The evidence on this point is consistent and clear, Einstein does not point to any facts of experience. He does, however, cite the success of earlier theoretical accounts of the propagation of light. For example, in *The Meaning of Relativity*, Einstein writes that,

because the Maxwell-Lorentz equations have proved their validity in the treatment of optical problems in moving bodies ... the consequence of the Maxwell-Lorentz equations that in a vacuum light is propagated with the velocity c, at least with respect to a definite inertial system [identified as "quiescent ether"] must therefore be regarded as proved.[29]

Again, in the important Appendix V added in June 1952 to his early 1917 *Relativity, The Special and General Theory*, Einstein writes

the special theory ... takes over from the theory of Maxwell-Lorentz [identified as "the theory of an ether at rest"] the assumption of the constancy of the velocity of light in empty space.[30]

We have dwelt on the evidence for this point because it is important to establish beyond reasonable doubt that, in 1905, Einstein was led to maintain that the one-way speed of light is constant and isotropic on the basis of a commitment to the *theoretical* results of Maxwell's and Lorentz's work on light and electrodynamics. Since the assumption of such properties for the one-way speed of light is equivalent to the assumption of standard signal synchrony, we may conclude that Einstein's choice of $\epsilon = \frac{1}{2}$ was not based on questions of conceptual or manipulative convenience, as the conventionalist account would suggest, but on the basis of prior theoretical commitments.

Within the conventionalist thesis, such a choice of standard signal synchrony would still be seen as a conventional choice, for the thesis maintains that there are infinitely many empirically equivalent ϵ-formulations of, for example, Maxwell's equations and that a commitment to the standard formulation can only be made on the grounds of convenience. There is, however, no evidence that Einstein was aware of this far-reaching claim of the conventionalists. Indeed there is little evidence that Einstein was aware of many of the important results of the conventionalist thesis. As far as we know, he never entertained the possibility of synchrony rules other than that of standard synchrony. This is an important matter since such speculation forms the core of the thesis. Einstein even writes,

... based on observations of double stars, the Dutch astronomer De Sitter was able to show that the velocity of propagation of light cannot depend on the velocity of motion of the body emitting the light. The assumption that this velocity of propagation is dependent on the direction "in space" is in itself improbable.[31]

The speed of light in question is clearly its one-way speed. Einstein's second remark here directly contradicts one of the fundamental claims of the conventionalist thesis that the one-way speed of light can indeed depend to a large measure on direction in space by arbitrary stipulation.

But what of Einstein's 1905 discussion of the need for a definition to underlie the concept of distant simultaneity? This discussion was repeated several times over the following years.[32] And further, what of Einstein's extreme care to claim in 1905 that only the two-way speed of light can be known *empirically* to be constant? Thus, in contrast to his *postulation* of the conjecture contained in the L-principle, Einstein is able to write,

In agreement with *experience* we further assume the quantity $2AB/(t_3 - t_1) = c$ to be a universal constant − the velocity of light in empty space[33] (our italics and notation).

Such a careful distinction can only result from an acute awareness of the fact that direct experiments, carried out in the nineteenth century, on the propagation of light rays *in vacuo*, such as those performed by Fizeau, Foucault and Michelson, had hitherto determined only an average speed of light over an out-and-return path, the two-way speed of light. Noting the perceptiveness of this distinction, can we not view the conventionalist thesis as a historical thesis which accurately portrays Einstein's approach to the concept to simultaneity?

We believe that too strong an emphasis on the conventionalist aspects of Einstein's 1905 paper distorts the contents of that discourse. Its focus is the work of Maxwell and Lorentz on electrodynamics. We have already seen that it was their work that led Einstein to choose standard signal synchrony. The main objective of the Kinematical Part of the paper was to recover that set of coordinate transformation equations, the Lorentz transformation, under which the fundamental equations of electromagnetism, Maxwell's equations, remain form-invariant. We can be certain that Einstein would have arrived at this result even before he embarked on his analysis of the nature of distant simultaneity. The importance, in 1905, of Einstein's analysis of simultaneity was that it enabled him to present a convincing derivation of the Lorentz transformation, one that was consistent with his strong positivist orientation. The main insights that Einstein had, in 1905, into the nature of time were not what we now see as the content of the conventionalist thesis, for, as we have argued above, Einstein did not carry through the results of his analysis of simultaneity to the same extent as modern conventionalists. Rather, Einstein's main insight was that the time coordinate used in the Lorentz transformation was not a theoretical construction derived from some more real absolute time, but that it was the only "real" time, the time indicated by clocks.

It is a mark of Einstein's acumen that he can so effectively draw together such apparently disparate methodological and theoretical concerns. A full appreciation of the nature of his 1905 work on special relativity must include consideration of both these aspects. The results of such an analysis are clear. His commitment to an empiricist methodology leads Einstein to view the definition of distant simultaneity as conventional but his equal commitment to the theory of Maxwell-Lorentz leads him to the no-choice situation of $\epsilon = \frac{1}{2}$.

4. LIGHT, LIGHT SIGNALS, SIMULTANEITY AND SPECIAL RELATIVITY

Light and light signals play a fundamental role in the conventionalist thesis and in nearly every formulation of Einsteinian relativity. We have seen earlier that, retrospectively, Einstein unequivocally accounted for his decision to feature light so prominently. He chose light because he believed that more was known for certain about light propagation than about any other process which could be considered for incorporation in the foundations of his theory.

We maintain that the full implications of this admission on Einstein's part have not been seen in many accounts of special relativity. In particular, we have in mind those traditional accounts which, following Einstein, use a discussion of the discovery of some rather unexpected properties of light to lead into and justify the construction of the theory.[34] Such accounts are misleading, for they leave one with an exaggerated impression of the importance of light in special relativity and the universe. One is led to believe that light has special significance even for phenomena which apparently have little or nothing to do with light, for many are governed by natural laws in which the speed c of light plays a fundamental role. Indeed, all such phenomena appear to conspire to prevent any causal process propagating at faster than the speed of light. This preferred position held by light seems assured by the fact that the new 1905 definition of distant simultaneity, the starting point of the whole enterprise, must use light signals and not some other, such as sound signals, and the fact that one of the two basic postulates of the theory is the light postulate, which, on a simple reading, consists of a statement about a property of light.

What is misleading in such an account arises not from what is said but from what is left unsaid. We maintain that these special properties, which appear to be attributable to light, are most properly seen as properties of space and time. Specifically we note here a property of space and time which has received much attention from Reichenbach and Salmon: that space will not allow signals to travel at greater than a certain maximum speed. We maintain that it is only because light is a first signal and travels at this maximum speed that it can feature so prominently in formulations of special relativity. What is misleading in the accounts of special relativity in question is that they do not proceed to show how the special properties of light arise from the nature of space and time. We shall soon turn to the task of doing just this, for it is the main purpose of this section.

First, however, we note that the origins of the problem lie in the strong phenomenalist epistemology which Einstein and others have coupled with relativity. Einstein first used this epistemology in 1905 to justify his elimination of the luminiferous aether from his theory and later, in 1916, to underpin his attempts to take away from space and time "the last remnant of physical objectivity".[35] From this standpoint, space and time, if they can be said to exist in any way at all, must be seen to be the most featureless of receptacles. Certainly, it would be impossible to attribute any active powers or properties to space and time. Now it is clear that, if such a viewpoint is adopted in an account of special relativity, it will be impossible for that account to treat adequately those features of the universe which are most properly seen as properties of space and time.

With the development of his general theory of relativity, Einstein came to write of the existence of inertial frames of reference, an active property of space and time within both classical mechanics and special relativity, as an "inherent epistemological defect" of both theories.[36] He saw his general theory as removing this defect and as the final step in his elimination of space and time as real entities.[37] In spite of this, in the same theory, Einstein found it necessary to attribute a very real and active property to space-time, that of curvature.

The empiricism of Reichenbach and Salmon is perhaps not as thoroughly phenomenalistic as that of the early Einstein when it comes to the question of space and time, yet they clearly continue to work within the tradition founded by Einstein. We see the concept of the first signal as a useful starting point for breaking out of this tradition. Although we take this concept from Reichenbach and Salmon, we do not take it without an important modification. Reichenbach and Salmon present the concept as a simple fact of experience: no signal can travel faster than a certain maximum speed, that of light.[38] We present this concept as arising directly as an active property of space and time: space will not allow any signal to propagate faster than a certain maximum speed. The difference here is crucial and it is essential if one is to understand that the apparently preferred position of light signals arises from the properties of space and time.

So far, in this paper, we have been content to use light and light signals in the traditional way, looking upon them, however, more as examples of first signals than as electromagnetic waves. We now turn to examine the status of light in the theory of relativity and the relationship between its properties and those of space and time.

In most traditional formulations of special relativity, the principle of relativity and the light postulate are presented as the axiomatic foundation of the theory. Now the principle of relativity is not a problem, in this context, for it manifestly is directly concerned with the properties of space and time. This is not the case with the light postulate, however, for this postulate seems to say more about light than about space and time. So we seek another postulate to replace the light postulate, one that will more clearly reflect our belief that special relativity is fundamentally concerned with the properties of space and time. We have seen from our earlier discussion of the conventionality of simultaneity that a key difference between the Newtonian universe and that of special relativity is that, in the latter, we are more isolated from distant events. Indeed, the thesis suggests that events separated in space in a relativistic world are isolated temporally in such a fundamental way that the very concept of the simultaneity of such distant events becomes problematic.

These considerations and others lead us to state our belief that the status of light in special relativity is more accurately reflected by replacing the light postulate in the foundations of the theory by the postulate that there exists a maximum speed at which signals can propagate in space. Or, stated more clearly as a postulate about the properties of space and time, that space will not allow any signal to propagate at a speed greater than a certain constant maximum. We shall denote this maximum speed, the speed of the first signal, by "c" and note that we have evidence that leads us to believe that its value is 3.00×10^8 ms⁻¹.

We can readily see that this postulate, in conjunction with the principle of relativity, is as powerful as the light postulate insofar as it is capable of yielding the familiar results of special relativity. For example, we note that according to the principle of relativity, the value of c will be the same in all inertial frames of reference. Next, we assume that

(i)   Space and time are homogeneous and isotropic,

(ii) the principle of relativity holds,

(iii) a signal travelling at velocity c in one inertial frame of reference also travels at velocity c in any other inertial frame of reference, and

(iv) distant simultaneity is determined through the rule of standard signal synchrony with $\epsilon = \frac{1}{2}$.

It is an elementary consequence that under these constraints the only co-ordinate transformation that can connect inertial frames of reference is the Lorentz transformation, with the proviso that we look upon the constant c in the transformation as the speed of a first signal rather than that of light.[39]

We now consider the question of the position of light in special relativity. To do this we do not look upon light as a first signal but initially as a phenomenon described by electromagnetic field theory. We shall first briefly examine field theories in the Newtonian universe and then look at the types of modifications that must be made to them to incorporate them into a relativistic universe.

We take as a Newtonian universe one in which the principle of relativity holds and there is *no* maximum to the speed of signals. In such a universe, transformations between inertial frames of reference are carried out by the so-called Galilean transformation equations.[40]

Within this universe, the most readily admissible field law is that which results from an inverse square force law, such as Newton's law of gravitation or Coulomb's electrostatic law. That such laws are the most readily admissible can be seen by writing them in a differential form. So, for example, Coulomb's law becomes

$$\nabla^2 \phi = -\frac{1}{\epsilon_0} \rho \qquad (1)$$

with the associated force law

$$\mathbf{F} = -\nabla \phi \qquad (1a)$$

where ϕ is the electrostatic potential, ρ charge density, ϵ the permittivity of free space and $\mathbf{F}$ the Coulomb force per unit charge.[41]

Now it is immediately clear that this field law remains form-invariant under the Galilean transformation. This is a result that we insist upon for any field law in the Newtonian universe as a direct result of the principle of relativity. It is for this reason that we do not admit the field law described by Maxwell's equations into the Newtonian world, even though historically this field law arose within the conceptual framework of the Newtonian universe. Indeed, attempts to incorporate this field law into the Newtonian universe led to the ultimate demise of that paradigmatic world view.

We now look at the types of changes that must be made to such inverse square law field theories when we attempt to incorporate them into a relativistic universe. In qualitative terms, two effects are of most interest. In the Newtonian version, any change in a field-producing charge will be reflected instantaneously by an immediate change in the distant field produced by that charge. Now this state of affairs is impossible in a relativistic universe, for it would correspond to a signal travelling at infinite speed through space. In a relativistic universe, we would expect that any change in a field would propagate through the field at a speed less than or equal to that of the first signal. In other words, the effect of the postulate of the maximum speed of a first signal leads us to expect the phenomenon of propagation of field perturbations in a relativistic universe.

The next modification is more subtle, but it is one that has the greatest ramifications for our ontology. In the Newtonian theory, momentum and energy conservation, when particles and fields interact, can always be maintained solely by considering the instantaneous interchange of energy and momentum between the particles alone. It is possible to side-step the question of the reality of the field, in the sense that it may exist independently of its sources and carry both energy and momentum, by giving an account of the interactions only in terms of action-at-a-distance forces between particles.

This, however, is no longer possible in a relativistic world. Consider a small charged particle falling towards a large charge under the action of the field of that charge within a relativistic universe. As it falls, the particle continuously gains energy and momentum. Now imagine that the source body is moved away from its original position to a position remote from the small particle. Since there is a maximum speed to signal propagation, the small particle cannot be affected until some finite time after the source body has moved away. During this time, the particle will continue to gain energy and momentum, even though the original source charge is no longer in proximity. What can the source of this action be? Surely it can only be the field that persists in the neighbourhood of the particle. What can the source of the energy and momentum be? The laws of conservation of energy and momentum require that they must have some source. Again, the source is the field itself.

This is a result of the greatest significance. We find, in a relativistic universe, that, as far as our mechanics is concerned, we must give the field an ontological status equivalent to that of sensible matter itself. Fields clearly have an existence independent of their source charges and, in particular, we must look upon them as carrying energy and momentum.[42] Combining this with our earlier result that, in a relativistic universe, we should expect to find perturbations propagating in any fields that it may contain, we conclude that such perturbations must carry energy and momentum, as for example, light waves are known to do.

All that remains to be done to conclude this account is to show that such perturbations propagate at the speed of a first signal and to give reasons for identifying them with light. To do this, we need to ask what types of changes need to be made to the Newtonian field law (1) for it to be incorporated into a relativistic universe. What we must do is alter its form so that it remains form-invariant under a Lorentz transformation, that is, Lorentz covariant.

We start by replacing the Laplacian operator ∇^2 on the left hand side of (1) by the Lorentz covariant d'Alembertian

$$\Box^2 = \nabla^2 - \frac{1}{c^2} \frac{\partial^2}{\partial t^2}$$

The potential ϕ can remain unaltered since as a scalar it is Lorentz-invariant. The source term on the right-hand side of (1), the charge density ρ, is not a Lorentz covariant quantity, given that quantity of charge is a Lorentz invariant. Based on this, however, we can readily construct Lorentz covariant source terms that are scalars, four-vectors or even tensors of higher order. We are led to select the Lorentz covariant four-vector (ρ, j), where j is the charge flux, as the source term, on the condition that the quantity of charge remain the measure of quantity of field source. But this now means that we have a scalar term on the left hand side of the field equation and a four-vector on the right as a source term. This suggests that we should extend the left hand side to a four-vector by looking upon the scalar potential as the timelike part of a four-vector, which we shall write as $(\phi, c^2 a)$. Thus the final form of our field equations is

$$\Box^2 \phi = -\frac{1}{\epsilon_0} \rho$$

$$\Box^2 A = -\frac{1}{c^2 \epsilon_0} j \tag{2}$$

Similar considerations lead us to augment the original force law (1a) with extra terms

$$F = -\nabla \phi - \frac{\partial A}{\partial t} + u \times [\nabla \times A] \tag{2a}$$

where u is the velocity of the charge upon which the force acts.

The import of this can be readily seen. First, we note from the properties of the d'Alembertian that we can expect the propagation of field perturbations to be wave-like *in vacuo*, and that these waves will propagate at the velocity c, which in our account is the speed of first signals. Second, we recognize (2) as Maxwell's equations and (2a) as the Lorentz force law, all expressed in terms of vector and scalar potentials, a and ϕ.[43] This suggests the identification of the relativised Coulomb field with the electromagnetic field and the identification of the waves we expect to find in it with electromagnetic waves — light waves. Finally this allows the setting of the speed of a first signal at the speed of propagation of light *in vacuo*.

We now summarise the results that have been obtained here. When a simple field law is translated from a Newtonian universe to a relativistic one, the existence of perturbations which propagate energy and momentum can be expected to arise in the field. In particular, the transference of Coulomb electrostatics from the Newtonian universe to a relativistic universe leads to the need for certain relativistic corrections to be added to the field law. These corrections turn out to comprise effects attributable to magnetic fields. Further, the perturbations that we expect to arise in such a field turn out to be wave-like, and they travel at the speed of a first signal. They are none other than light.

Now the crucial point is that all that separates a Newtonian universe from a relativistic one is a simple but important property of space and time. In a Newtonian universe, there is no upper limit to the speed at which a signal may propagate, whereas in the relativistic universe there is such a limit. (In our account, the principle of relativity holds in both Newtonian and relativistic universes.)

Thus we regard the speed at which light travels and its invariance over all inertial frames of reference as being determined by the properties of space and time, as opposed to being an inherent property of light, or, more generally, of the field in which it arises. We even look upon the very fact that light exists as stemming ultimately from the properties of space and time. More specifically, we can say that the phenomenon of light arises automatically when we place the simplest of field laws, that of the Coulomb field, into a universe in which the field law is to be form-invariant over all inertial frames and there is a maximum speed at which signals can propagate.

Suppose we now ask, why is it that light seems to play such a prominent role in relativity; why is the speed of light an invariant over all inertial frames; why does it appear in so many laws of nature which have nothing to do with light *per se*; why is it an upper limit to the velocity of all causal processes? It can now be seen that these questions are misplaced. Once it is recognised that the constant c is a fundamental constant arising from the nature of space and time, such questions cease to pose a problem. It then becomes clear that one should not ask, for example, why it is that the speed of light is the maximum speed at which signals can propagate. Rather, one should ask why it is that light propagates at the maximum speed that space allows for signal propagation. The answer to this question, we believe, is found in the analysis of field theory, in a relativistic universe, given above.

We conclude this account by returning to the question of the use of light signals in definitions of simultaneity and the use of the light postulate as a key part of the foundations of relativity. We stress that there is no serious quarrel with formulations of special relativity that make prominent use of these devices, for one is at liberty to set up the theoretical structure of special relativity using whatever axioms and definitions one sees fit, provided the axioms and definitions are consistent. Now we have shown that light is a first signal. So this condition is certainly true of the light postulate and the definitions of simultaneity based on light signals. Indeed there are many circumstances in which such formulations have distinct advantages. For example, the use of the light postulate in the foundations of the theory provides a direct and simple connection between empirical findings on the nature of space and time, through experiments on light propagation carried out in the last century, and the axiomatic foundations of the theory. For this reason, it was highly appropriate for Einstein to choose this postulate as part of the foundation of the new theory in 1905.

It is our contention here, however, that such formulations of the theory are most properly used only in conjunction with a clear understanding of the position that light holds in the theory, along the lines that we have set out above.

5. CONCLUSION

In our reconstruction of the genesis of Einstein's 1905 formulation of the "definition of simultaneity", we hope not to have misconstrued the reasons for his commitment both to the conventionality of distant simultaneity and to the choice $\epsilon = \frac{1}{2}$, standard signal synchrony. Provided one recognises that Einstein did not follow through the implications of his own analysis of simultaneity to the same extent as the modern conventionalists, there is no epistemological difficulty in holding both that distant simultaneity is conventional, in the sense that theories embracing different values of ϵ are empirically equivalent, and that in 1905 Einstein nevertheless had good reason for the assumption that $\epsilon = \frac{1}{2}$, through his commitment to the Maxwell-Lorentz theory.

We have also drawn attention to one of the consequences of Einstein's coupling of special relativity and a narrowly phenomenalistic epistemology which insists upon the elimination of space and time as entities. Varying degrees of commitment to this epistemology have hindered the emergence of an understanding of why light signals in particular can be used to define distant simultaneity and why the constant c, numerically equal to the speed of light *in vacuo*, should feature so prominently in the laws of nature. We believe that such understanding arises when one sees space and time as entities possessing active properties. Then it becomes clear that the special properties of light, and even its very existence, arise from the nature of space and time themselves.

The lesson that we should draw from this is surely that there may well be significant gains to be made from a reappraisal of the epistemologies that underlie the theory of relativity and its associated writings.

APPENDIX

Considerable stress has been laid in this account on the need for looking upon the constant c as a property of space and time, rather than of light. We have treated c as the maximum speed at which space will allow signals to propagate, since this has been most suitable for our account. There is, however, an alternative, and possibly more recent interpretation of the relationship between c and space and time which will be discussed here because it highlights the fact that c is a property of space and time.

It is well known that one of the basic concepts of general relativity is that Euclidean geometry is no longer adequate as an empirical theory accounting for the properties of real space. This result was, however, foreshadowed by the results of special relativity. One of the results of special relativity is that it is impossible to have a rigid rod in real space. This strikes at the heart of Euclidean geometry, looked upon as an empirical theory of the properties of space, for to use the theory as such one needs to stipulate a correspondence between the theoretical entity, the line interval of Euclidean geometry, and an entity in real space, the rigid rod. But special relativity denies the existence of such rods, making such a stipulation impossible.

We can look on the single most important insight provided by the theory as the following: We must extend the domain of the empirical science of geometry to include kinematics, the study of motion, if the theory is to retain maximum generality. We do this by considering the joint study of the geometry of space and of kinematics as encompassed by the study of geometry of four dimensional space-time. Many interesting complications arise when we try to do this. One is that we find that the formula for the metric, the formula that gives us the "distance" between two point-events in space-time, is only pseudoeuclidean.

The result that is of considerable interest to us here arises from the question of the commensurability of displacements measured along the three space-like axes with those measured along the time-like axis. These measurements are not directly commensurable, since one will be made in space units, e.g., metres or feet, whilst the other will be made in time units, e.g., seconds or years. That these measurements must be somehow rendered commensurable is an essential prerequisite of our constructing any serious four-dimensional geometry of space-time. This is achieved by introducing a conversion factor to enable us to relate time unit and space unit measurements, a factor which in some aspects is much the same as the conversion factor of 100 used to relate measurements in centimetres to those in metres. This factor turns out, in the theory, to be c. We could hardly ask that c correspond to a more fundamental quantity in the theory of space and time.

This makes it clear that to look upon c solely as the speed of light is to approach this important constant on the most elementary of phenomenological levels.

University of New South Wales (Sydney)

NOTES

1 Cited by Seeling [1956], p. 75.
2 See, for example, Salmon [1969b], [1975a], Chap. 4, and [1977].
3 Salmon [1977k], p. 254.
4 Einstein [1952], p. 40 in conjunction with Scribner [1963].
5 Reichenbach [1927], pp. 126–127.
6 Reichenbach [1927], p. 127.
7 Winnie [1970].
8 Salmon [1977k], pp. 270–271.
9 Ellis and Bowman [1967].
10 Winnie [1970].
11 Salmon [1969b].
12 Salmon [1977k], pp. 270–287.
13 Whittaker [1949], p. 58.
14 Reichenbach [1927], p. 143.
15 Grünbaum [1955].
16 Salmon [1977f], p. 217.
17 Salmon [1975a], p. 122.
18 Salmon [1977k], p. 268.
19 Winnie [1970].
20 In a paper that has aroused much interest, David Malament has claimed that the conventionalists are fundamentally mistaken and derives the result that the only admissible synchronisation rule is that of standard signal synchrony. It is important to keep the context of his result in mind. It is derived from an axiomatisation of the properties of space-time which is based on the relation of causal connectibility between events. Within this system, light signals are taken as the fundamental "measuring stick" of space-time. Thus it is not surprising that the system does not admit nonstandard synchronisms, for these are equivalent to an anisotropy in the one-way propagation of light. Such anisotropy cannot even be defined in such a system, for this would require yet another independent "measuring stick". Malament's result is, however, very serious for accounts of the conventionalist thesis in which the causal theory of time is invoked but not for those in which clocks or rods are taken as primitives for the measurement of time and space. See Malament [1977].
21 Poincaré [1898].
22 Poincaré [1902], p. 90.
23 Einstein [1916] and also Einstein [1949], p. 53.

[24] Einstein [1921], p. 27 and Bowman [1976], p. 74.

[25] Einstein [1948].

[26] Einstein [1952], p. 38.

[27] Einstein [1952], p. 41.

[28] Einstein [1952], p. 38.

[29] Einstein [1921], pp. 25–26.

[30] Einstein [1917], p. 148; see also Einstein [1907] and [1936].

[31] Einstein [1917], p. 17.

[32] See, for example, Einstein [1911], pp. 7–8 and [1917], pp. 21–24.

[33] Einstein [1952], p. 40.

[34] See, for example, Einstein's own [1917].

[35] Einstein [1952], p. 117.

[36] Einstein [1952], p. 112.

[37] Einstein [1952], pp. 109–164.

[38] Reichenbach [1927], pp. 143 and Salmon [1975a], pp. 105.

[39] We note here in passing that as an added bonus this formulation of the theory eliminates *ex hypothesi* the possibility of the existence of tachyons, postulated faster than light particles. Their incorporation into special relativity has been problematic for, apart from introducing imaginary quantities whose physical interpretation is far from clear, they also lead to the possibility of closed causal loops. See Salmon [1975a], p. 122.

[40] Grünbaum [1955], p. 10.

[41] Panofsy and Phillips [1972], pp. 8–11.

[42] Of course, unlike sensible matter, the energy-momentum of the field has zero rest mass. This does not mean that the field should have a lower ontological status in the theory. Rather its main significance is to require that the energy-momentum of the field always be in motion at the speed of a first signal.

[43] Panofsky and Phillips [1972], Chap. 18.

GRAHAM NERLICH

SIMULTANEITY AND CONVENTION IN SPECIAL RELATIVITY

1. INTRODUCTION

What physics books are apt to say about SR (Special Relativity) is not quite the same as what philosophy books are apt to say about it, as Wesley Salmon points out in his excellent *Space, Time and Motion* (Salmon [1975a], p. 113). He explains this difference reasonably enough, as due to disparate main interests which SR has for physicists as against philosophers. The former want to develop quickly an apparatus which allows the clear, deft portrayal of central principles and results in physical prediction and explanation. The latter prefer a more leisured approach to this goal so as to give scope for a deeper insight into the semantic-syntactic structure of SR. Most philosophy books say that the language of SR has various conventional elements in it, which means that the theory can have no very simple relation between its syntax and its semantics. In particular, the matter of the simultaneity of space-like separated events is settled conventionally and this gives rise to a contrast in SR between sentences which form a factual core (Winnie [1970], p. 229, Salmon [1975a], p. 117) and others which make up a periphery of non-factual sentences with a merely syntactic function. In what follows I ignore the problem of what other conventions might have a place in SR. I want to examine and reject just this idea that simultaneity is a convention, as this gives rise to the idea that we can contrast a core of factual sentences of SR with a periphery of merely conventional ones.

Wesley Salmon's thought on this central problem is certainly conventionalist. Besides its admirable clarity and precision, Salmon's work contains many valuable arguments and observations on the structure of SR. He has illuminatingly discussed a wide range of "alternative methods" of synchrony (Salmon [1977k]). He has proposed a useful qualitative measure of conventional triviality (Salmon [1969b]). In his book (Salmon [1975a]) he makes explicit the very important distinction between the question whether simultaneity is a ternary empirical *relation* among two events and a frame and the question whether it is established by *convention* in a frame; he also gives an informative and provocative account there of the significance of superluminally fast particles, known as tachyons. Like many others, I have

Robert McLaughlin (ed.), What? Where? When? Why?, 129–153.

learned much from his (and other) scholarly and elegant discussions of conventionalism to which I am deeply indebted in my own attempt to state a quite different view of the status of simultaneity in SR. If it is discovered that, in learning something I have still not learned enough, the mistakes will be all my own.

What do physics textbooks tell us about SR? The best of the more recent ones tell us that it is a theory about a 4-dimensional physical manifold, spacetime, with Minkowskian metric in which we can describe a wide range of physical quantities mainly in terms of 4-vectors and tensors. In this, light propagation fills a distinguished place since electromagnetic wave fronts (in a vacuum) occupy what are called the null geodesics of spacetime. Time-like geodesics describe the trajectories of force-free particles. I assume, in common with many others, that spacetime physics, framed in concepts of the style I have just been illustrating, gives the best account of SR.

Physics books sometimes say other things, using other concepts. They tell us, for example, that whether or not two space-like separated events occur at the *same time* is a relation, not just between the two events but among the events and a *frame of reference*. They tell us that SR is special in being confined to a restricted class of frames of reference in each of which the laws of physics take the same, invariant form and in each of which the speed of light (in a vacuum) is a universal constant c. It is on this latter way of picturing SR that philosophy books claim to improve. They say that, given a pair of space-like separated events and a frame of reference, it is not a relational fact but a convention whether the events are simultaneous. This means that, given both the events and the frame, we can still freely choose to say that the events are simultaneous or that they are not simultaneous. Neither choice states (or misstates) a fact and neither sentence chosen is apt for semantic appraisal nor has a truth value. We simply determine the language by a decision on the matter, the decision is conventional, and we gain no factual content for SR by making it.[1]

This picture of things began with Einstein who wrote (in 1905):

we establish *by definition* that the 'time' required by light to travel from A to B equals the 'time' it requires to travel from B to A (Einstein [1923], p. 40. Italics in original. The translation quoted is a correction; see Grünbaum [1973], p. 344, fn. 4).

It is hard to overestimate the impact of this remark on philosophy in this century, set, as the words were, in the context of a highly successful, revolutionary theory of physics. Though Einstein used tools already forged by logical empiricism there is no doubt that his employment of them in this brilliant paper immeasurably enriched their apparent range and power. It is still a methodological orthodoxy in the philosophy of space and time to see many apparently real questions as calling for answer by factually arbitrary, conventional stipulations. The conviction that simultaneity questions, in particular, call for such an answer is still the central stronghold in the empire of conventionalism.

The strategy of my arguments will be clearer, I hope, if I make two comments on Einstein's predicament in 1905. Firstly, he faced a major problem in teaching us (and himself) how it is *conceivable* that simultaneity might be relative to a frame. He had to undermine what appeared to be the synthetic *a priori* truth that it is absolute. It is not surprising that he should have looked to a positivistic epistemology to solve the problem. I will argue that we can see, in hindsight, that the problem was not epistemological and therefore that the solution was misconceived. Secondly, Einstein wrote before Minkowski discovered how to find the description of the spacetime world which lay encoded in Einstein's description in the 1905 version of SR. I shall argue that, on the one hand, a spacetime perspective enables us to see most clearly what Einstein's simultaneity problem really was, despite the fact that, on the other hand, questions of simultaneity can properly be raised and settled only in a classical ontology in which space and time, as separate entities, do *not* 'fade away into mere shadows' (Minkowski in Einstein [1923], p. 75). At the same time we must provide a basis for translating, without loss of content, SR physics in its spacetime interpretation into SR physics in its space and time interpretation. A condition of adequacy which constrains the concept of a frame of reference is that *it* should provide a basis for these complete translations. Frames of reference, as conceived of by Einstein in 1905, and as generally conceived of in philosophy since, do not meet this adequacy condition, a fact which Einstein could not possibly have understood before Minkowski's great step forward to spacetime. The idea of a *convention* for simultaneity arises out of these confusions. It is to be rejected by clarifying them and cannot, I think, be shown mistaken by propounding various styles of experimental means for fixing simultaneity.

2. COORDINATE SYSTEMS FOR SPACETIME

I follow many others in seeing SR as primarily a theory of spacetime rather than of frames of reference for space and time. This is important since spacetime affords no natural foothold for the ideas of frame of reference, speed of signals, simultaneity of events. It is conceptually incongruous with the question conventionalism aims to settle. I shall spend a paragraph making this not unfamiliar point more explicit.

I call the language in which we speak about spacetime and spacetime objects a 4-language and the ontology of this language a 4-ontology. Now we may not say, in 4-language, that spacetime objects move in spacetime nor that they remain at rest in it. Nothing in spacetime structure entitles us to say of two time-like separated points that they are 'at' the same spatial place or 'at' different places at different times. 'Same (spatial) place' is not part of 4-language. Similarly we can neither affirm nor deny of two space-like separated points that they are 'at' the same time, or simultaneous. We cannot say of two time-slices of a spacetime object that 'it' is the same thing (at the same place again). 'It' is neither a (continuant) thing nor the same. Thus, in 4-language (and in 4-ontology, therefore) there are no (continuant) things which move or remain at rest and nothing makes or unmakes the occurrence of two events be at the same or at different times. Nothing is to be said, in 4-language, about the speed of any thing, not even about light's speed being (or failing to be) constant nor being the greatest speed. Light rays have a privileged status, indeed, for they lie in the cone of null geodesics which is metrically determinate at every spacetime point and distinct from every non-null geodesic of the manifold. In 4-language we can attribute invariant properties to spacetime objects; the spacetime interval between spacetime points, the shape and size of material hypervolumes in spacetime can be described invariantly, no matter by what varying coordinate differences we specify them in different systems of coordinates. Our 4-language is not a (continuant) thing language; it says nothing about a (continuant) space nor about (spatially global) times. Spacetime physics invites the use of a quite particular array of concepts to describe spacetime and its objects and claims to give an exhaustive account of these. In this array of concepts the idea of a frame of reference, in particular, has no place. I will defend this last claim in more detail in Section 3.

For describing spacetime and its physical contents it is convenient to choose a system of coordinates. This is not at all the same thing as choosing a frame of reference — not in the conventionalist's book, quite certainly, and not in mine, as will become clear. It is a truism to say that coordinates are conventional. This means at least two things: first, the quadruple of coordinate numbers ascribes no property to the point which has them in the way that a quadruple of numbers might describe an energy-momentum or electromagnetic vector, a spacetime object, a that point; second, and more importantly, we can describe the vectors, tensors, the metric of spacetime and so on without the use of coordinates, if we wish to. A coordinate system is just a global device of reference, a (dispensable) means of representation, nothing more — save for the proviso that we expect coordinate differences to reflect the measure of the interval along coordinate lines.

Another way to bring out the fact that coordinates are conventions of representation is to point out that we expect our spacetime physics to be formulated in a *generally covariant* way. That is, although the coordinate components of, say, a vector vary from one arbitrary system of coordinates to another we expect that the components will change according to the same transformation as allows us to go from the one set of coordinates to the other. The components are covariant with the coordinates. This effectively robs the coordinates of any significance beyond that of representation. It tells us, for example, that the privileged status of light as a null geodesic depends not at all on any system of coordinates which may be in use. (But, of course, this privileged status does not ascribe a *speed* to light, as a signal.) That SR can be expressed in a generally covariant way has been argued for in a conventionalist's paper (Giannoni [1978]) and is not a matter of dispute.

I now want to make two points which seem to me very obvious ones indeed in the light of what has been said. Evidently all that a question of "simultaneity" of two events could come to *in a coordinate system* with axes x_0, x_1, x_2, x_3 is the trivial matter whether or not the system gives them the same x_0 coordinate. Though this is indeed trivial and reflects no fact about the events and their posture in spacetime it would be quite misleading to describe it as a *convention within the system* whether they have the same coordinate or not. The system itself is a convention; we can get along without any such system. That we have chosen this system rather than another is a convention, if you like, since no fact about spacetime is reflected in choosing it. But given the system there is no *further* matter to be settled — by convention or by anything else — as to whether the events have the same or different x_0 coordinates. That is simply a *relation* they bear to the coordinate system in use and is fixed in choosing it. Precisely parallel remarks apply to the idea of 'same place' in a coordinate system. It can come to no more than the trivial matter whether the system assigns two events the same space coordinate. But, of course, no one ever did think *that* was a convention. It was always clearly understood as relational.

Next, we need four coordinates if we are to give an adequate *coordinate* description of four-dimensional spacetime and its physical contents. That is to say, that part of the system of coordinates which assigns the same (or different) x_0 or time coordinate to space-like separated events is no less necessary for a complete coordinate description than the parts which assign the same (or different) x_1, x_2, x_3 or space coordinates to them. Without all four coordinates for any pair of points we cannot express, for example, the invariant spacetime interval between the points in our coordinate language. We cannot define which geodesics are null by coordinate means. Each coordinate is quite as necessary as any other for the coordinate expression of the facts which SR has to tell us about the structure of spacetime and its contents.

Compare the coordinate geometry of the two-dimensional spatial plane, confining ourselves to linear, but not necessarily orthogonal, coordinates. Let us suppose that someone notes our freedom to choose any of the lines of the plane as an x-axis. Consequently, he says that x coordinates of (and x coordinate differences between) points in the plane are *relative* to a choice of x axis. He also notes our freedom to choose any of the lines of the plane (not parallel to the one chosen as x axis) to be a y axis. However, he describes choice of a y axis not as completing the choice of a relatum suitable for the complete coordinate description of geometrical objects in the plane, but rather as a *convention* adopted *within* the chosen relatum of an x axis. He asserts that it is a question of relational fact whether a line is or is not parallel to the x axis, thus giving all the points on it the same y coordinate (no matter which y axis we choose). Therefore, a sentence which states this factual relationship (of parallelism) has a truth value. However, he claims, it is something quite different, a convention, whether a line is or is not parallel to the y axis, thus giving all the points on it the same x coordinate (whichever x axis we might have chosen) so that a sentence which states *this* parallelism has no truth value and corresponds to no state of affairs. Further, he tells us, there are no facts about y coordinate cross-sections of figures generally in the plane unless their sides are parallel to the x axis. Of course, there are facts about the dimensions of these figures relative to *new* x axes, chosen so as to parallel the sides. But relative to a given x axis the y-cross-section geometry of figures in the plane generally is no question of fact and can be settled only by the free conventional choice of y axis. This, I suggest, would be a deeply misleading account of the nature of coordinate geometry in the plane.[2]

I suggest that if anyone were to adopt a conventionalist picture of the relation of time-like to space-like axes for the coordinate geometry of spacetime it would be misleading a quite similar way, Of course, the situation is more complex in spacetime. In particular, we can define invariant intervals of spacetime without recourse to space-like axes, treating the geometry in what approaches a coordinate-free way. We can use a measure of interval along time-like lines (proper time), a 'rest length' metric which 'spaces out' parallel time-like lines (but is emphatically *not* a measure of interval along space-like coordinate lines) and the 'light ratio' which connects these metrics. This is given by what is, essentially, the familiar two-way light principle. Given two points a and b on a time-like line A it relates the proper time (measure of the interval) $b - a$ to the 'rest-length' metric from A to any of the time-like lines parallel to it which contain just one point of the (space-like) sphere given by the intersection of the light cones through a and b. The figure (which 'doubles up' those usually drawn to represent the two way light principle) makes this graphic, I hope.

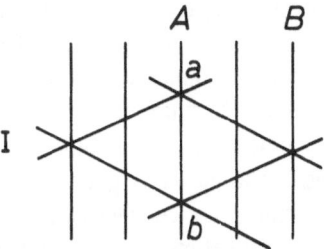

$(b - a)|$ measures proper time *along* A from a to b. $(B - A)$ counts the number of unit-parallels from B to A. It is *not* a measure along any space-like curve from A to B.

What Winnie's ϵ-variable treatment (Winnie [1970]) shows is that provided we can measure intervals along the curves A we do not *also* need a measure along space-like curves (such as spatial coordinate axes for spacetime would provide). The 'rest length metric' and the light ratio suffice. I will call this a *pre-coordinate* geometry. What this means, in diagram II is that the 'rest length' tells us that the time-like lines A B C are, say, one rest-length metre apart without defining how this relates to the measures of the various space-like intervals between A, B and C taken along the lines 1, 2, 3.

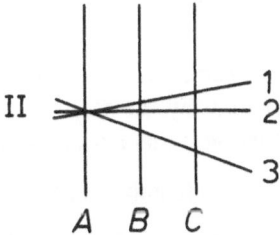

No doubt this is deeply significant. But it does not even *appear* to deny that we *can* do full coordinate geometry if we wish. Nor does it deny that, if we do, then the selection of space-like axes closes all questions whether or not two events have the same time coordinates. Nor, again, does it seem to deny that the selection fixes a perfectly definite space-like geometry for the coordinate time slices of 4-objects whether or not their time-like extended lines are parallel to our time-like axis, i.e., whether or not they have coordinate velocity. There are certainly factual matters about the size and shape of *all* such time-slices of 4-objects, and their inclination or lack of it to the chosen time axis is simply irrelevant. In fact, we can deduce precisely what these spatial properties are *from* Winnie's ε-treatment of SR which suffices to define the measure of *any* interval of spacetime, whether space-like or time-like. That we *need* not do coordinate geometry does not mean that when we do we find some sort of *factual slack* to be filled out by convention.

Next, it would be a mistake to suppose that we have here come across a unique geometric role for time-axes and for time-like lines generally. We can turn diagram I on its side, make perfectly good sense of it and have it yield a route to the spacetime metric no less elegant than before. For simplicity, consider a two dimensional spacetime. Let *a* and *b* be space-like separated points in spacetime and let *c* be the point of intersection of the (upward) light cones through *a* and *b*. If we select *a* and *b* as points defining a space-like axis then a question might be said to arise whether *c* occurs at the mid-point (*m*) between *a* and *b*. Suppose someone asserts that, relative to this space-like axis it is a convention whether *c* occurs at the same place as *m*; that this can be settled only by stipulation to

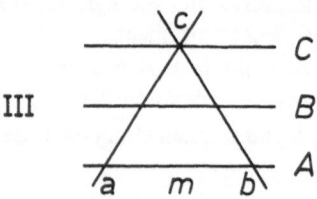

III

the effect that light from *a* towards *b* goes the same distance as it goes from *b* towards *a* in the time spaced out by the space-like parallels *A*, *B*, *C* between *ab* and *c*. The simplest stipulation is that *c* occurs at the coordinate place

$$c = a + \frac{1}{2}(b - a) \qquad (\text{compare } t_2 = t_1 + \frac{1}{2}(t_3 - t_1))$$

but so far as the physical facts are concerned, we get exactly the same physics by replacing $\frac{1}{2}$ by any other real number in the open interval $(0, 1)$. To avoid arbitrary, non-factual conventions or assumptions, we must use the expression

$$c = a + \epsilon(b - a) \qquad \text{(compare } t_2 = t_1 + \epsilon(t_3 - t_1)\text{)}.$$

Now provided we have a proper length (measure along the space-like curve A from a to b), a metric for simultaneity intervals which 'time out' the space-like parallel lines A, B, C, and the light ratio, then we can get along without a time-like axis. We can construct the geometry of space-time in a pre-coordinate way, defining the invariant spacetime intervals analogously to our (to Winnie's) procedures before. Briefly, let me show just how light-like paths define null geodesics independently of choice ϵ. The space-time interval $a\, c$ is given, in ϵ-notation thus:

$$\Delta x_\epsilon^2 - c_\epsilon^2 \Delta t^2 \qquad \text{(compare } \Delta x^2 - c_\epsilon^2 \Delta t_\epsilon^2\text{)}$$

that is

$$(\epsilon(b - a))^2 - \left(\frac{(\epsilon(b - a))}{\Delta t}\right)^2 \Delta t^2 = 0$$

$$\text{(compare } \Delta x^2 - \left(\frac{\Delta x}{\epsilon(t_3 - t_1)}\right)^2 (\epsilon(t_3 - t_1))^2 = 0\text{)}.$$

I conclude that this new ϵ-free-variable treatment has all the advantages of Winnie's treatment and shows just as much (and as little) about what is conventional and what relational in SR.

Admittedly there is no simple means of measuring the interval along a space-like track analogous to the way a clock measures a time-like one. But this is beside any point about which sentences are factual and which conventional. First, we can measure the interval operationally by taking the (unique) maximum of the rest length measures between all pairs of parallel time-like lines which include points a and b respectively. This assumes nothing about simultaneity nor does it need the passing of signals. There is no way to reject the identity between this maximum of the rest length metrics and the proper length along the space-like interval while keeping intact the links which bind the metric of spacetime to natural measures of space and time. Again, Winnie's ϵ-variable treatment allows us to deduce the proper length (and, indeed, had better do so on pain of inadequacy) since it is a spacetime invariant. It is worth noticing that the pair of parallels which gives us this maximum is, in fact, orthogonal to the space-like line a, b, thus parallel to the time-like axis of any Lorentz coordinate system which takes the line as its space-like axis. This appears already to provide us with a signal- and transport-free criterion for preferring $\epsilon = \frac{1}{2}$.

There may indeed be much interest in pursuing SR without assumptions as to the distance-in-one-direction-per-unit-time with respect to space-like frames of reference. But it seems to me quite gratuitous, indeed actively confusing, to regard rest or motion not as relative but as conventional for this reason. All the expressions, in my attempt to paraphrase Winnie's treatment, which concern *same place, rest, motion*, and so on are quite incongruous with what is under discussion and cloud a clear insight into what it is about. Questions of rest and motion, like those of simultaneity, have no place in spacetime geometry.

Despite the much greater complexity of spacetime geometry then, it does seem that a conventionalist style of treatment of coordinate geometry of spacetime is no less misleading than the conventionalist account of the coordinate geometry of the Euclidean plane, which I parodied before. It is true that choosing a time-axis leaves us perfectly free to choose spatial axes from among any of the spacelike lines of spacetime. But it is equally true that choosing a set of three spatial axes leaves us quite free in the choice of time axis. General covariance guarantees that. There is no priority among axes here despite the fact that we may not rotate any timelike axis into a spacelike axis (though we may rotate any spacelike axis into any other). From the standpoint of spacetime physics one finds no reason at all for singling out x_0 or time axes for special status. They appear to have no sort of superior philosophical interest, nor do spatial axes create a special foothold for the use of conventions within coordinate systems. Different choices of axes in no way disturb the privileged status of the light cone as composed of null geodesics, nor do they affect in the slightest the invariant intervals of spacetime itself. In fact, from this viewpoint it seems impossible to discover anything which conventionalist theories of simultaneity shed the least light upon.

3. FRAMES OF REFERENCE

However, these are merely manoeuvres preliminary to fixing on the area where I take conventionalism to be properly debatable. Conventionalists argue not about coordinate systems but about frames of reference. My next objective is to make clear what conventionalists have meant frames of reference to be and that much of the difficulty about simultaneity has been created by this idea's being inadequate and unfruitful as conventionalists apply it. It is clear enough that some epistemological advantages make philosophers' frames of reference seem important, but I will argue that the less said about them (and expressed by means of them) the better. Unless, of course, we give up talking about the sorts of frames of reference philosophers usually deal with, and follow the usage of physicists by adopting a fundamentally different concept of frame of reference.

A frame of reference is something which picks out of spacetime a space and a time. This is rough and a bit misleading, as we will see, but it does suggest at once why a frame of reference is not just a coordinate system. A frame of reference is insensitive to many criteria which distinguish coordinate systems, such as the difference between polar and Cartesian coordinates for space, a rotation of spatial axes and a translation of the origin in space or in spacetime. There is a 1-1 correspondence between frames of reference and *classes* (uncountably large classes) of coordinate systems. Members of one of these classes are alike in respects more deeply significant than those which distinguish one member from another. What significance is this? It is the importance of a point of transition from one array of concepts for dealing with physics to another array; it is the significance of a bridge towards an alternative language and ontology.

A frame of reference only *corresponds* to a class of coordinate systems. For example each continuous set of spacetime points all with the same x_0 (time) coordinate in some system of coordinates picks out a space-like hypersurface in spacetime. But none of these hypersurfaces is a space since the space-like hypersurface $x_0 = t$ is *not identical* with the hypersurface $x_0 = t'$, and no point in the one hypersurface is the *same place* as any point in the other. Unlike any coordinate system, a frame of reference enables us to speak of space, of time, of continuant things with a three-dimensional shape and size which endure in time, of rest (the *same* place at *different* times), of motion, of speeds and velocities, and, lastly, of simultaneity (the *same* time at *different* places). These, like the term 'frame of reference' occur not in 4-language but in another language, with other conceptual structures and with another ontology. It is a space and time, or a 3 + 1-language (and ontology). The idea of a frame of reference is much older and more familiar than that of a spacetime coordinate system, deriving from classical, post-Newtonian mechanics. It gives point to a number of questions which have only trivial counterparts in the ontology of spacetime. To be specific, it gives them an ontic or structural point. In 3 + 1-language we can pointfully ask whether the speed of light is the same in all directions, what the (spatial) shape and size of an object is, whether it is moving or not, which way and how fast. Indeed, the syntactic structure of the language *requires* that we ask these things.

It is unclear how seriously we can or should take the metaphysical view that the 3 + 1-physics is no trivial alternative ontology for SR. But unless it is a significant alternative then questions of simultaneity, the speed of light and the rest are not significant either. That the ontologies differ only trivially would be no ground for recommending conventionalism but rather for seeing matters in the light of the observations on spacetime coordinate systems made in Section 2. These make no sort of sense of 'conventions of simultaneity.' Unless 3 + 1-physics provides a significantly different point for questions which have no point in 4-physics then conventionalism has nothing to offer but a distinctly confused account of spacetime. I incline rather strongly towards the view that the ontic shift is significant (though *my* inclination is not really relevant to the arguments of this paper). The *motivation* for a 3 + 1-ontology is clear enough. It provides the concepts under which SR is epistemically accessible to us. However, the conventionalist doctrine is about the factual content of SR as contrasted with its linguistic content. It claims that SR has a factual core and a merely linguistic periphery. (See, e.g., Salmon [1975a], p. 117. Nerlich [1979] expands this core-periphery picture of conventionalism and criticises it in connection with affine and metric properties.)

To make clear how a 3 + 1-ontology might give importance to questions which have no corresponding point in 4-ontology I will look at an example which I think begs no question against *e*-conventionalism. Consider the problem of the relativity of motion. In classical physics we cannot take as at rest, frames accelerated or rotating relative to inertial frames since this creates, quite artificially, inertial, centrifugal or Coriolis forces of which we can give no proper account. We would be obliged to posit *causes* yet there are none. However, if we consider generally covariant formulations of SR[3], then we can formulate our 4-physics with respect to arbitrary curvilinear coordinate systems. If we now treat frames of reference as trivial equivalents of these coordinate systems, then we get in some sense, a physics in which the previously unacceptable accelerated or rotating frames are recognised as proper. Of course, we get the artificial inertial, centrifugal and Coriolis forces exactly as we do in classical physics. There is no better explanation available for them than there was before, yet it seems no less pertinent to ask for their causes and sources provided that we take the 3 + 1-ontology seriously. Now, however, we simply wave these objections away as trivial since we refuse to recognise a significant difference between coordinate systems and frames. In effect, *we reject Newton's ontology* as a serious alternative to the new 4-ontology but continue to pay lip service to classical language in the insignificant forms of translation just described. (Nerlich [1976], Ch. 10, Section 10 amplifies this style of argument.)

If the idea of a frame of reference is to lend some significance to the debate of whether simultaneity is a convention as against a relational fact, then a frame of reference ought to be a bridge between the ontology and ideology (array of concepts) congruous with spacetime and the ontology and ideology congruous with a (continuant) space which endures in time. At least two criteria for the adequacy of candidates for frames of reference seem to be indicated by this:

(1) A frame of reference maximises the class of well-formed sentences in 3 + 1-language which can be semantically appraised (state facts) in terms of structures having well-defined counterparts in 4-ontology (and vice-versa).

(2) There should be no *facts* stateable in 3 + 1-language which call for (causal) explanations which the resulting 3 + 1-physics cannot provide.

These criteria are loosely stated but it would be laborious to make them more precise and they will serve my purpose. They beg no question against conventionalism as far as I can see. What are candidate correspondences between reference frames and classes of coordinate systems?

Let us define an *L-frame of reference (or linear frame)* as a class of linear spacetime coordinate systems equivalent save for translations of origin in spacetime and for rotation of any x_i coordinate about any x_j $(i, j \neq 0)$. In the correspondence, an L-frame obliterates much that serves to distinguish coordinate systems (position of origin, orientation of spatial axes one to another) but it is sensitive to which partitioning into a space and a time is brought about by a given coordinate system. It disallows rotation of spatial coordinates about the temporal and vice versa. The class of coordinate systems gives us a family of parallel time-like lines, each having constant x_1, x_2, x_3 coordinates in any member of the equivalence class, and each having a definite proper-time metric along it. The class also gives us a definite family of space-like hypersurfaces, each point in any hypersurface having constant x_0 coordinates in any coordinate system in the equivalence class and each having a definite (Euclidean) metric. Thus, for each point event in spacetime an L-frame provides a corresponding (spatial) place of the event at other times and a corresponding time of the event at other places.

The Lorentz frames of the physicists are one, but only one, kind of linear frame. In the present section it is not my purpose to consider whether Lorentz frames form a preferred class of linear frames (which is, more or less, the question whether $\frac{1}{2}$ is a preferred value for ϵ). That is the topic of Section 5. My present aim is to contrast L-frames with the (philosophers') T-frames so as to consider the question of which matters in SR are factual and which conventional (which is, more or less, the question what is convention and what fact when we assign ϵ *any* value we like in the open interval $(0, 1)$). I shall argue that to say there are conventions at issue is to misdescribe the way in which 3 + 1-sentences correspond to 4-sentences.

If our main requirement on a reference frame is that it should bridge the two ontologies so that we can find the same physics in both, then L-frames go a long way towards meeting it. Sentences in the language of the 4-ontology correspond quite unambiguously to sentences in the other language, and vice versa. Each L-frame has a continuant space and we can speak of the same place at different times so that description of the relative inclination of timelike lines in spacetime corresponds in a definite way to a description of rest or motion of particles. The space of any frame has a well-defined Euclidean geometry and each continuant object of the 3 + 1-ontology has a definite shape, size, relativistic mass and so on whenever the counterpart object in the 4-ontology has well-defined counterpart properties. Further, if we look into any of the (infinitely many) coordinate systems which are members of the frame we do not thereby discover new properties fixed by these systems which we would be obliged to recognise as physically significant in the 3 + 1-ontology.

If we see frames of reference in this ontological light, then they are not conventional in the same sense as coordinate systems are despite the close connection between coordinate systems and frames. We can do 4-ontology physics without coordinate systems but we cannot do 3 + 1-physics without frames of reference, since statements about the facts of rest and motion cannot be made at all except *in relation* to some frame or other. There is no 'frame-free' 3 + 1-physics, however free we may find ourselves to choose among frames. Further, we would certainly expect an adequate bridge between the 4 and 3 + 1-languages to provide for an adequate semantic appraisal of all the properties which the syntactic structure of 3 + 1-language provides for things. For example, there are going to be well-formed sentences attributing a definite metric shape and size to every object. If, under any proposed correspondence between the languages, the well-formed sentences cannot be semantically appraised then, on the face of it, that is a reproach to the correspondence rather than an indication that we are to eke out the correspondence by arbitrary assignment of "truth" values to certain of these sentences. Certainly if another form of correspondence *does* fully provide for the semantic appraisal of the sentences that surely means, not that *conventions* are appropriate in the proposed correspondence, but rather that it is an *inappropriate form of correspondence*.

Let us focus, now, on the main problem we are pursuing, as it applies to
L-frames. L-frames define *same place*. Do they define *same time* or simultaneity relative to an L-frame? Quite obviously, they do. Given an L-frame,
there is no room whatever for a convention to settle whether events e_1 and e_2
are simultaneous: there are no syntactically different, semantically equivalent
'redescriptions', no alternative choices of ϵ. Simultaneity is relative to a frame,
however, there being no absolute space or time common to *all* such frames.
That simultaneity is well-defined for a linear frame is the secret of this object's
success in bridging the two ontologies.

Plainly, this is not the idea of a frame of reference which we find in phi-
losphy books.

What, then, is the conventionalist's concept of frame? It is the idea of a
T-frame (or timelike frame). A T-frame corresponds to a class of linear spacetime coordinate systems equivalent save for translation of origin in spacetime
and for rotation of any x_i coordinate axis about any x_j axis ($i = 1, 2, 3; j = 0,$
1, 2, 3). T-frames, then, correspond to classes of linear coordinate systems
equivalent just in that their x_1, x_2, x_3 = constant timelike trajectories yield
the same family of parallel timelike lines. A T-frame does not define a partitioning of spacetime into spacelike hypersurfaces. Each T-frame contains *all*
the families of spacelike hypersurfaces into which we might partition spacetime in the sense that it is the set of all affine coordinate systems with the
same time axis, no matter how the spacelike axes of these various systems
may be disposed. In the diagram, each of the coordinate systems Tx, Tx', Tx'',
$T'x$, $T'x'$, $T'x''$ is a member of a different L-frame from the others. But Tx,
Tx', Tx'' are all members of the same T-frame which is defined just by T,
whereas $T'x$, $T'x'$, $T'x''$ are all members of the different T-frame, defined by
T',

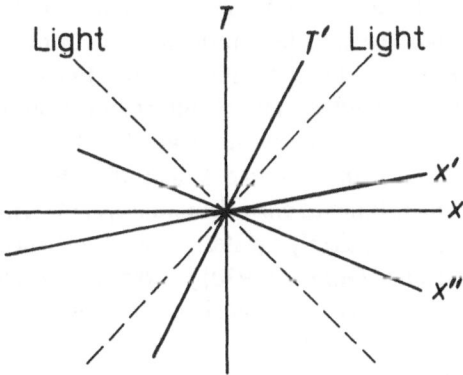

More illuminatingly, perhaps, a T-frame is best understood not as the $3 + 1$-counterpart of a class of coordinate systems for spacetime but rather as the counterpart of what I called a pre-coordinate system (in Section 2). As we saw, this treatment of spacetime is not really a coordinate geometry description, though certainly a no less satisfactory one. The question whether T-frames provide a suitable basis for a correspondence between $3 + 1$- and 4-languages need not be at all the same as the question whether pre-coordinate procedures for describing spacetime geometry are suitable. We already saw that there cannot be a frame-free treatment of $3 + 1$-physics.

It is reasonably clear that those who have debated the issue of conventionalism in simultaneity have intended T-frames to be frames of reference. After all, L-frames obviously do not permit a convention whereas it might seem that T-frames do. What do we find in the literature? Frequently the phrase 'frame of reference' is used without any gloss at all. Einstein spoke, in 1905 of "a 'stationary' coordinate system" (Einstein [1923], p. 38) and he is followed in this by Reichenbach, who also speaks of the relativity of *motion of bodies*. (Reichenbach [1958], Section 34, esp. p. 217). This suggests that a body or set of rest bodies would constitute a frame, which would be equivalent to specifying a T-frame. Of course one finds many discussions of the distinction between inertial and non-inertial frames of reference, but these focus on questions of dynamics and very seldom venture anything which clarifies explicitly which *spacetime* objects are under discussion. All of this permits the T-frame interpretation, though one could wish for something more specific. Janis (Janis [1969], p. 74) defines a frame of reference as a family of time-like curves with spatial coordinates attached and a metric along these curves fixed by proper time intervals. That seems quite clearly to be a pre-coordinate system and plainly corresponds to a T-frame. So far as I can discover only van Fraassen uses the phrase to refer to a coordinate system for spacetime. This puts him in the position of claiming that a frame of reference is *constituted* by conventions.[4] (van Fraassen [1969], p. 68). It allows him to argue, in effect, that simultaneity is among the conventions which constitute a frame but it does not allow him to say (nor does he say) that it is a convention within a frame. I think that this difference between van Fraassen and others is not of great significance, in the last resort. Giannoni (in his [1978], pp. 22–25) clearly equates frames of reference with T-frames.[5] I take Winnie's ϵ *variable* (Winnie [1970]) as ranging over all the L-frames in a T-frame. This crucial observation has already been made (not in quite these terms) in Friedman [1977], p. 419.

It is quite clear that we cannot say, of any pair of spacelike separated events, either that they are or that they are not simultaneous relative to a T-frame. However, it would be a confusion to think, therefore, that simultaneity needs to be fixed *by convention* in a T-frame. Simultaneity sentences are conceptually incongruous with T-frames as relata. More clearly and explicitly, T-frames fail to provide a basis by means of which well-formed sentences necessarily included in the syntactic resources of 3 + 1-language can be determinately matched with sentences in 4-language. Sentences about simultaneity are only one of the kinds of sentences which T-frames do not allow us to map. Therefore, T-frames fail a main requirement placed by SR on the idea of a frame of reference. This may not be obvious. After all, we can speak of the same place at different times, that is, of rest and motion in a T-frame. Further, each T-frame appears to identify a continuant space which has a definite (Euclidean) geometry. More, objects at rest have well-defined physical properties (shape, size, temperature) each of which corresponds to a physical property of its 4-dimensional counterpart. But objects in motion relative to a T-frame cannot be said to have any of these latter properties. A T-frame provides a determinate mapping for (sentences expressing) these properties only up to affine versions of them, so that we cannot even specify their size relative to one another or to rest objects.[6] Yet the corresponding (coordinate) properties of the 4-ontology counterparts of these objects are no less well defined than they are for the 3 + 1-objects at rest relative to the T-frame.[7] In fact, a whole range of Lorentz-variant properties of 3 + 1-objects are ill-defined in a T-frame for any object in motion relative to the frame. However there surely must *be* definite properties of these moving 3 + 1-entities. If not, the counterpart 4-entities would be indefinite, too. T-frames in short do not allow us to construct the correspondence needed to provide a semantic role for an important class of 3 + 1-sentences, though a semantic role *can* be provided by using L-frames as a bridge. So, using T-frames as a bridge, the 3 + 1-theory says less than it might and it does not map the 4-language counterpart facts fully. It is perfectly clear what to conclude. *Choice of a T-frame does not allow us to state the full physical content of spacetime physics in the language of a 3 + 1-ontology.*

There appears to be something like agreement among conventionalists that to choose a T-frame leaves a need of some sort for a specification of ϵ. But is this a convention? First, choice of ϵ *adds to the class of sentences in 3 + 1-language which can be appraised semantically*. That is, it increases the number of well-formed sentences in 3 + 1-language which correspond to facts which have well-defined counterparts in structures of 4-ontology. I do not see how this can fail to mean that it adds to the factual content of the corresponding 3 + 1-language. Choice of ϵ does not *result* in there being well-formed 3 + 1-sentences which lack truth value, state no fact, have no semantic role in SR. *Exactly* the opposite is true. Well-formed sentences which are *required* (by syntactic structure) to state facts if the 3 + 1-language is to correspond everywhere to 4-language are not yet provided with a fact to state, with a basis for semantic appraisal, so long as ϵ is unspecified. In particular, facts about space-like coordinate cross-sections of objects whose time-like lines are not parallel to the time-axis are just as real and objective, just as determinate, as facts about the cross-sections of objects with no coordinate velocity. To put the point a little differently, if we discriminate more finely among the coordinate systems of those classes each of which corresponds to a T-frame we *do* thereby discover new structures defined by these systems which provide facts for 3 + 1-sentences to correspond to. Hence the virtue of L-frames.

I am arguing then that the *full coordinate geometry* of spacetime forms the most satisfactory basis on the side of 4-ontology for a SR correspondence with 3 + 1-ontology and that, therefore, an L-frame forms the most satisfactory basis on the 3 + 1 side. In that case, simultaneity, like rest, is simply relative to a frame of reference of the kind appropriate to 3 + 1-physics. The correspondence between an L-frame and a pre-coordinate description of spacetime is not satisfactory. This is reflected in the emergence of the ϵ-problem and the uncertain doctrine of conventionalism which follows it. The need for L-frames rather than T-frames is not well-reflected in the view that ϵ ought to be specified as a convention though it does not matter how. Nor do I think it illuminates the matter to regard the conventionalist theory of ϵ-choice as equivalent to providing a freely chosen *further term* for a relational fact.[8] First, as was argued in Section 2, we can use pre-coordinate methods for spacetime just as effectively through *space-like* lines, planes, and hyperplanes as through time-like lines. It would be no less an error here than in spacetime to see the structure of SR conferring superior ontic status on time-like as against space-like elements. Second, it seems superfluous to see the need for proper semantic appraisal of 3 + 1-sentences as calling for *two* relata, a T-frame and ϵ, when the concept of an L-frame as a *single* relatum is so natural and so readily defined.

One might look to find an asymmetry which favours T-frames over L-frames in the epistemology of SR which, after all, is what motivates serious attention to 3 + 1-ontology. There can be no doubt that our experience makes the distinction of rest and uniform motion in a frame much more accessible to us than is the shape of a rapidly moving body. This is a large problem on which I shall make only three brief comments. First, I take ϵ-conventionalism not to be an epistemologically based theory but to be concerned with the structure of the real world as SR describes it. Secondly, though the facts of rest and motion are more directly accessible epistemologically than facts about the space-like cross sections of bodies inclined (in spacetime) towards our x-axis this by no means implies that the latter are inaccessible. They are in fact deducible quite unambiguously by the pre-coordinate methods discussed in Section 2. Lastly, SR neither states nor presupposes that the world fails to provide us with directly accessible, naturally occurring space-like hypersurfaces. In fact, the world does not provide them. But suppose that a continuous 3-dimensional subset of spacetime points were all to show a particular feature (radiation of a certain frequency is emitted from each, say). If this subset of points constitutes a space-like hypersurface it would certainly be quite directly observable that it did. There might be regularly 'spaced out' series of these, each member parallel to the others (analogous to parallel time-like lines), members of any one series being inclined in spacetime to members of others. Then the world would present us with choices between simultaneity classes just as accessible observationally, just as intelligible and just as purely relative as it now presents in respect of classes of rest points. So far as I understand what might be involved in this (and this proviso might be important) I do not believe that SR tells us *anything at all* about why the world is not so structured. The physics of SR is at least as much a physics of fields as it is of particles — in fact there are no strict particles in reality — and nothing about its structure, expressed either in 4-language or 3 + 1-language, provides a necessarily superior ontic or structural status to time-like lines as against space-like ones.

4. HISTORICAL REMARKS ON THE CONCEPT OF A FRAME OF REFERENCE

A very brief historical sketch clarifies, at least to some extent, why T-frames have assumed an importance which they really do not have. It also helps to make it intelligible why a disproportionate emphasis has been placed on the claim that the speed of light is the *fastest* signal and I hope to use this material to show how the conventionalist picture of SR leads to misunderstanding of the theory.

Newton understood motion as the relation of a body to an absolute (unique) continuant space. It was motion *through* space. He conceded that there might be no body at all in a state of rest, i.e., remaining in the same place in this

unique space, and he also pointed out (roughly following Galileo) that a state
of rest was not distinguishable by experiment from uniform linear motion
through space. This led later physicists to regard it as pointless to attempt
the discovery of Newton's unique space and to construct frameworks suitable
for the conduct of physics by letting the laws themselves delete 'rotating'
or 'accelerating' frames in favour of inertial ones. Had it been true that choice
of a framework could not influence which classes of events are simultaneous
it would have been true that choice of a suitable framework was equivalent
to choosing some body or bodies as at rest.[9] From the point of view of a
Newtonian spacetime, choosing rest points is equivalent to choosing a family
of parallel timelike lines, but since the stacking of spatial hypersurfaces is
unique up to sliding and shearing transformations (affine transformations)
such a framework does provide a basis for translating 4-sentences of classical
physics into 3 + 1-sentences without loss of content, without rendering well-
formed sentences impossible to appraise semantically.

In the light of this, it seems correct to say that, in 1905, Einstein's real
discovery was that the laws of physics take the same form under Lorentz
transformations from any one *suitable basis* for physical description to any
other *suitable basis*.[10] Part of this discovery is that not all speeds vary under
such transformations but that a certain finite speed (namely the speed of
electromagnetic propagation in a vacuum) is an invariant of the transformations. This is how Minkowski described the discovery in 1908 when he gave
it a spacetime interpretation.[11] It asks us to recognise that the classical means
of identifying a frame of reference, by choosing a family of rest points, no
longer provides a suitable basis for 3 + 1-physical description. I am arguing
that Einstein misunderstood this aspect of his discovery and that later conventionalists have persisted in the misunderstanding.

The account just given is, I argue, more accurate than the account which
tells us that his main discovery was that light is the first or fastest signal. The
latter account suggests both that Einstein's epistemological style of treatment
was proper, that many 3 + 1-sentences of SR physics have no truth value, and
that admitting 'particles' moving relative to some proper frame of reference at
'speeds' greater than light would undermine SR. Had this latter account
been correct, then Einstein's discovery would have been of even more fundamental importance than it was, yet very difficult indeed to assimilate into
physics. It would have been the discovery that objects in motion are *indefinite*,
that there is no matter of fact about what shape or size they have and that
though we attribute shapes, sizes, and much else to them in our 3 + 1-physics,
such sentences merely are conventionsl, have no truth value, state no fact, are

not suitable objects of semantic appraisal. However, this lack of definition would, so to speak, be movable from one set of objects to others as we change our frame of reference, in a way which seems distinctly bizarre. If this were the structure of the world it is not clear either why we should pretend (by adopting an arbitrary convention) that moving objects are definite in ways in which they are actually indefinite, why we should want to reify, on this non-factual basis, the whole new ontology of spacetime, or alternatively, how we could understand a physics in which the content of perfectly factual sentences in the language of 4-ontology becomes fugitive, conventional and indeterminate in the allegedly equivalent language of 3 + 1-ontology.[12] Surely it is clear that once we take seriously a 3 + 1-ontology, then there *must*, in fact, be definite shapes, sizes and so on which moving bodies have just as much as rest ones do and that this demands a simultaneity relation which is itself a *factual relation* on which all these other factual relations are dependent.

This completes the longest part of my argument. Let me state, again, what I think I have established. The major error in conventionalism has been to attempt to give an account of SR based on an inadequate 3 + 1-concept of a frame of reference. Unhappily most opponents of conventionalism have paid no closer attention to frames than the conventionalists themselves so that the status of those sentences left, so to speak, with no means of semantic support has been obscured. I hope that my view of their status is now quite clear. Unless every last one of them states a fact then the content of 3 + 1-theory is not equivalent to that of 4-theory. This certainly does not point to the need for a convention which retains these sentences without giving them a semantic role. On the contrary, it points to the inadequacy of philosophers' frames of reference as a basis for complete description of the facts which SR tells us make up the world. The first lesson is: drop T-frames and use L-frames instead. The second lesson is: relative to L-frames sentences about simultaneity, the speed of light and the rest are factual, informative about physical reality, true or false. That is the pivot round which the rest of this paper turns, if it turns at all.

5. THE PRIVILEGED STATUS OF LORENTZ FRAMES

A crucial problem still lies before us. Lorentz frames are a proper subset of L-frames. Are they a privileged subset or can we equally well choose any linear frame for our 3 + 1-physics? From a more physical point of view this might look like the very same question as the one we raised at first: can we choose any value for ϵ in the open interval $(0, 1)$? However, on a hopeful

view of the arguments of Sections 2–3 it has now been stripped of its conventionalist associations and is seen as a question about the *relativity* of
simultaneity among *L*-frames just as surely as questions about rest and motion
are not conventional, but relational. We need only bear in mind that simultaneity, the shape and size of moving objects are relational, factual questions
to see that the facts which lead us to prefer Lorentz frames are quite obvious
and simple ones.

First, there are the facts about the slow transport of clocks, already sufficiently discussed in Ellis and Bowman [1967], Ellis [1971] and Friedman
[1977].

Second, as Malament has shown (Malament [1977]), the class of Lorentz
frames is uniquely definable in spacetime in terms of the concepts of time-like,
light-like and space-like connectibility.[13] This is a result of very considerable
importance, I believe. Nevertheless one might appeal against it on behalf of
L-frames by pointing out that, if spacetime really is an entity in its own right,
then there can be no preferred *coordinate system* (or set of systems) which
is to be used in describing physical events in that manifold. This simply
enunciates a principle of general covariance for real physical entities in a
real spacetime. However, as was argued earlier (pp. 141–2), questions which
have no point in *4*-language may well, necessarily, have point in 3 + 1-language.
Our second adequacy condition for a frame of reference, stated there, was as
follows:

> There should be no facts stateable in 3 + 1-language which call for
> (causal) explanations which the resulting 3 + 1-physics cannot
> provide.

We can exploit this to provide other grounds for viewing Lorentz frames as
privileged.

There is a quite clear and simple reason for preferring Lorentz frames,
since only these provide us with an isotropic space for the frame. Clearly,
space will not be isotropic for the propagation of light in any frame where
$\epsilon \neq \frac{1}{2}$. But not only this is the case. The length of a moving object, the rate
of a moving clock, relativistic mass, and so on will all be functions, not just
of speed, but also of direction of motion in any *L*-frame where $\epsilon \neq \frac{1}{2}$. In fact,
once again, for a wide range of physical properties of 3 + 1-objects which
vary under Lorentz transformation, there will be dependence not just on
speed, but on direction of motion. Consequently there will be facts stateable
in 3 + 1-language which call loudly for causal explanation, *once they are
recognised as* (relational) *facts*. They would require causal explanation for

reasons quite similar to those which lead us to require explanation of Coriolis forces etc. and to reject certain (rotating) frames of reference which allow for no such explanations. See Nerlich [1976], Ch. 10, Section 10, esp. p. 257. The resulting 3 + 1-physics certainly cannot explain them.

Just such a criterion for preferring Lorentz frames was discussed by Grünbaum (in his [1973], p. 355). I have not found any later discussion of it. His dismissal of the criterion relies heavily on the conventionalist philosophy which the arguments of Sections 2–3 were directed against.

The contention that either the isotropy of space or Occam's "razor" is relevant here is profoundly in error, and its advocacy arises from a failure of understand the import of Einstein's statement that "we establish *by definition* that the 'time' required by light to travel from A to B equals the 'time' it requires to travel from B to A." For, in the first place, since no statement concerning a one-way transit time or one-way velocity derives its meaning from mere facts but also requires a prior *stipulation* of the criterion of clock synchronization, a choice of $\epsilon \neq \frac{1}{2}$, which renders the transit times (velocities) of light in opposite directions *un*equal, cannot possibly conflict with such physical isotropies and symmetries as prevail *independently* of our descriptive conventions. (loc cit.)

Grünbaum is surely correct in arguing that *if* one accepts that simultaneity is not a *relation* to a frame of reference but a *convention* within it, then one must concede that further apparent facts which would determine simultaneity must be conventional, not relational, too. In that case, the isotropy of space would be a convention, not a relational fact. But, I have tried to point out, simultaneity *is* relational, not conventional so the anisotropy of the space in the frame is real, not merely embodied in a sentence which lacks truth value, states no fact, cannot be appraised semantically. The criterion is formulated for the 3 + 1-ontology, not for spacetime; it yields a non-arbitrary relational simultaneity but not an absolute one. Since the criterion meets the conditions of adquacy argued for earlier, I conclude that Lorentz frames are to be preferred in SR, not just on grounds of simplicity, but because they alone achieve a partitioning of spacetime into times and isotropic relational spaces.

Once we allow the simultaneity of spacelike separated events to be a matter of fact relative to a suitable frame (i.e., a relational fact), the physical *evidence* to support standard signal synchrony as physically correct is truly massive. It, uniquely, makes space isotropic. It makes the lengths of rods, the rates of clocks, relative mass and so on, functions of a motion's speed but not direction. It gives light the same speed in all directions. It is verified by the slow transport of clocks. It is not easy to think of other matters of relational fact for which one can assemble so impressive a body of confirmation.

This completes my arguments for the conclusion that Lorentz frames give, uniquely, a factually satisfactory picture of SR in 3 + 1-ontology.[14]

University of Adelaide

NOTES

[1] I will make use of Reichenbach's ϵ notation. This has become familiar, but I shall briefly introduce it. Suppose there are two spatially separated points A and B in some reference frame. At t_1 on a clock at A, A sends a light signal to B which arrives there at t_2 on B's clock, is reflected back and arrives at t_3 on A's clock. Then t_2 (at B) is related to t_1 and t_3 (at A) by the formula:

$$t_2 = t_1 + \epsilon \, (t_3 - t_1), \ 0 < \epsilon < 1$$

That simultaneity is conventional is equivalent to the choice of ϵ's being conventional.

[2] It seems to me no improvement to say that whether two points have the same y coordinate may be equivalently described either as a *convention within* the relatum of the chosen x axis or as a relation both to the x-axis and to the y-axis. This appears to me to parallel Giannoni's account of the conventionalist's description of things. See Giannoni [1978], pp. 39–40.

[3] I mean by this that we deal with flat spacetime but place no restrictions on coordinate systems that they be linear. In some usages this formulation is regarded as the beginning of General Relativity, whereas in others this is seen as beginning only with the transition to spacetimes not necessarily flat.

[4] He is not quite so explicit as this might suggest but, on my understanding, is firmly committed to the claim attributed to him here.

[5] Incidentally, Giannoni (like many others) speaks of 'velocities for light . . . along the x axis' (p. 20) without reminding the reader that this is not at all the same thing as the x axis of any *spacetime* coordinate system. Perhaps this is obvious but I, at least, would feel that I understood conventionalism more clearly if the distinction between the axis of the *frame* and the x axis of a *coordinate system* were always sharply drawn. It would be strictly nonsensical to speak (a) of light's *moving* (b) as along the *space-like* line of an x coordinate axis of spacetime.

[6] Salmon [1975a], pp. 87–89 gives a striking example of the kind of indefiniteness I am speaking of. Whether the moving train is wholly inside the tunnel or overlaps it on either side is indeterminate in a T-frame.

[7] Though certainly well-defined, they are not defined in quite the same way. The length of a rest object is defined in spacetime by the 'rest length' spacing out of time-like parallels whereas the length of a moving object can only be defined as the length of a *space-like interval* of spacetime. But this difference is not a difference in any *determinacy* of factual structure. Both lengths can be expressed in the full coordinate geometry of spacetime and the pre-coordinate approach discussed in Section 2.

[8] That is, to equate for example the claim *that 'e$_1$ is simultaneous with e$_2$ relative* to F' *is not true but conventional* with the claim *that 'e$_1$ is simultaneous with e$_2$ relative*

to F and to $\epsilon = \frac{1}{3}$, *is factually true.* This is quite explicitly Giannoni's view. See Giannoni [1978], pp. 39–40.

[9] This presupposes that there are likely to be bodies which can be taken as at rest in inertial frames. However, it is not likely at all, and the presupposition has made the relativity of motion look much more like a principle which a positivist could embrace with a clear epistemological conscience than it has any right to look.

[10] This statement anticipates (and depends on) later arguments (Section 5) that, among linear frames of reference, only Lorentz frames are suitable. At present, I have argued only that *T*-frames are *not* suitable.

[11] Minkowski's lecture is reprinted in translation in Einstein [1923]. The remarks in this paragraph are drawn from pp. 78–79 of the lecture as reprinted there.

[12] In the event of this last outcome we would surely be obliged to conclude that 4-ontology provides, not just a more perspicuous, but the only possible ontology for SR.

[13] Malament says (p. 293) that this definition is provided in terms of causality, but I suggest that the description given here is more satisfactory in view of my claim that time-like lines have no privileged status in SR.

[14] I am indebted to Chris Mortensen for his careful discussions with me about this paper.

WESLEY C. SALMON

COMETS, POLLEN AND DREAMS:
SOME REFLECTIONS ON SCIENTIFIC EXPLANATION[1]

Now we know
The sharply veering ways of comets, once
A source of dread, nor longer do we quail
Beneath appearances of bearded stars.
                        Edmund Halley[2]

1. INTRODUCTION

The Newtonian synthesis, which provided the basis for all of classical physics, produced far-reaching changes in our ways of looking at the world. Laplace, who made significant contributions to the development of classical physics, was one of its most eloquent champions. Like Halley, he found the Newtonian explanation of comets an inspiring example of the power and value of modern science.

Despite the fact that classical physics still has wide applicability to various sorts of phenomena, we no longer believe it to be literally true. Nevertheless, it seems to me, certain philosophical views concerning the nature of science which arise directly out of a Laplacian conception of the world continue to exert an enormous influence upon current thought about scientific explanation. A caveat should be issued at once. I shall *not* be arguing the historical thesis that Laplace's writings had a direct influence upon contemporary philosophers; instead, I shall maintain that the general viewpoint which was expressed by Laplace, and which pervaded much of nineteenth-century thought, has carried over into the twentieth century and permeates much of contemporary philosophy of science.

Laplace was, of course, a firm advocate of mechanistic determinism; accordingly, he believed that biological phenomena and human behavior are as rigidly determined by the laws of Newtonian mechanics as are the motions of comets and atoms. Only our lack of knowledge prevents us from seeing that fact. Many nineteenth-century scientists, in the biological and social sciences as well as the physical sciences — steeped in the tradition of classical physics — believed that all of the phenomena in the world can ultimately, in principle, be reduced to classical physics. It was, I think, this

Robert McLaughlin (ed.), What? Where? When? Why?, 155–178.
Copyright © 1982 by D. Reidel Publishing Company.

Laplacian conception of mechanical determination, bolstered by at least a century of additional spectacular success of classical physics, that provided the model for scientific explanation most widely accepted by philosophers and scientists in the *twentieth* century. Stated succinctly, the claim is that, *with the aid of suitable initial conditions, an event is explained by subsuming it under one or more laws of nature.* This is hardly more than a translation into more up-to-date terminology of Laplace's colorful statement:

Given for one instant an intelligence which could comprehend all of the forces by which nature is animated and the respective situation of the beings who compose it – an intelligence sufficiently vast to submit these data to analysis – it would embrace in the same formula the movements of the greatest bodies of the universe and those of the lightest atom; for it, nothing would be uncertain and the future, as the past, would be present to its eyes.[3]

Such an intelligence, Laplace must have believed, would exemplify the highest degree of scientific understanding, and would be able to provide a complete scientific explanation of any occurrence whatsoever.[4]

In the closing months of the nineteenth century, Planck provided the basic building block – the quantum of action – of a new science that would undermine and supersede classical physics. Neither Planck, nor anyone else at that time, could foresee the fundamental conceptual revolution physics was destined to experience in the first quarter of this century. By 1926, Heisenberg and Schrödinger had formulated the basic theory of quantum mechanics, and Born had furnished the statistical interpretation, which has subsequently become the standard physical interpretation of the quantum theory. These developments, to say the very least, cast serious doubt upon the whole conception of Laplacian determinism. Physical science has, by now, fairly well absorbed the shock of supposing that the physical world may be fundamentally and irreducibly statistical, though some physicists still staunchly resist this interpretation of quantum mechanics. It is not clear, however, that philosophy of science – as expounded by scientists as well as philosophers – has digested this development, along with its repercussions for such concepts as scientific explanation. It was not until 1962 – an astonishing delay – that *any* systematic attempt was made to explicate statistical explanation, and it appears to me that the resulting analysis was far from satisfactory.[5] I suspect that too close an adherence to the Laplacian ideal may have been responsible for some of the difficulties.

In addition to considering statistical explanations, we shall find it necessary in the course of the discussion to take a careful look at functional explanations. Scientific progress has, rightly I believe, tended to purge science of teleological principles. Aristotelian physics, in which nature *abhorred* a vacuum and bodies *sought* their natural places, has been totally superseded by the mechanical physics of Newton.[6] Darwinian evolution, with its principle of natural selection, has replaced the doctrine that species were specially created by God to fulfill divine *purposes*. This laudable attempt to remove purposive and anthropomorphic explanatory principles from science has, I think, made many scientists and philosophers wary of functional explanations, and has encouraged the notion that in fully mature sciences functional explanations are eliminated in favour of other types. This has led some philosophers to characterize functional explanations as "explanation sketches" or "incomplete explanations".[7] Although it can be shown quite clearly, I believe, that certain types of functional explanation need not involve any anthropomorphic or teleological elements, philosophers and scientists have not universally been convinced of their scientific legitimacy. Nevertheless, pious hopes to the contrary notwithstanding, important classes of explanations in some sciences are functional explanations, and they are by no means patently reducible to explanations of any other type.

The purpose of the present paper is to re-examine the nature of scientific explanation from the standpoint of contemporary science. I shall pay careful attention to our heritage of Laplacian determinism – with its obvious bearing upon scientific explanation – but I shall also try to see how these conceptions have to be modified in the light of more recent developments. As my foregoing remarks have indicated, I shall devote considerable attention to statistical and functional patterns of explanation. In so doing, I shall be raising issues which are matters for consideration by scientists in a wide range of fields, from anthropology to zoology – touching psychology, quantum physics, and sociology, among others, along the way.

2. LAPLACIAN EXPLANATION (COMETS)

2.0. *Misconceptions*

An important benefit of Newton's explanation of comets was to render them less terrifying. This result is achieved, it has sometimes been suggested, by transforming the unfamiliar into something familiar.[8] Describing a comet as a planet-like object with a highly eccentric orbit does help to classify it with better-known objects and this, it is claimed, is what makes it more understandable.

Appealing as it may seem, this conception of explanation can hardly be considered adequate. It is easy to cite many examples in which the opposite occurs; the familiar is explained by invoking highly esoteric considerations. The outstanding instance is the Olbers paradox — why is the sky dark at night? No fact could be more familiar than the darkness of night, but any adequate explanation of that phenomenon will involve intricate cosmological considerations. Another familiar fact is that offspring resemble their parents in certain respects; its explanation takes us into the chemistry of the DNA molecule and the "genetic code". A third example is Freud's explanation of dreams — familiar occurrences to most people — in terms of unconscious wishes, which at the time (if not now) were unfamiliar to the point of being far-fetched. I do not mean to assert that the Freudian explanation of dreams is correct, but the fact that it attempts to explain the familiar by means of the unfamiliar is no obstacle to its acceptability.

A closely related notion requires that explanations must make ultimate reference to conscious aims and purposes, if they are to provide *genuine understanding*. Such explanations are teleological. We are familiar with the motives which explain many of our own actions; the demand is sometimes made that any explanation of any other phenomenon must refer to the purposes of the Creator of the world, or perhaps to some purpose which is inherent in nature itself. This view probably lies at the heart of the claim, often made in earlier times, that science in and of itself can provide only description — not explanation. Such a view of explanation has been severely criticized for its blatant anthropomorphism, and I doubt that it enjoys much support among contemporary scientists. At the same time, those sciences such as biology, sociology, and anthropology, which seem to make extensive use of functional explanations, have sometimes encountered serious problems in showing that they were not *ipso facto* involved in teleology. As I remarked above, I think that careful analysis can draw a viable distinction between those functional explanations which are teleological and those which are not. But it remains to be seen what role, if any, functional explanations can play in the overall scheme of scientific explanation.

Having considered some common misconceptions of the nature of scientific explanation, let us attempt to arrive at more adequate formulations. The plural "formulations" is quite deliberate and very important. I shall offer three characterizations of Laplacian explanation which, in that context, may seem to differ only terminologically. When, however, we move on to consider modifications of the Laplacian view demanded by developments in twentieth-century science, the differences take on crucial logical importance.

2.1. *The Epistemic Conception of Scientific Explanation*

Suppose that we attempt to explain a particular occurrence, such as a lunar eclipse, by citing certain laws which, together with suitable antecedent conditions, entail that the eclipse occurred at a particular time.[9] In this case we can plausibly say that the explanation is a valid deductive argument, with premises consisting of law-statements along with other statements which describe the initial conditions, and with the explanandum-statement as its conclusion. This explanation could be described as an argument to the effect that the event to be explained was to be expected by virtue of the explanatory facts. I shall refer to this view as the *epistemic conception* of scientific explanation. Given an event which, when it occurred, might or might not have been expected, we explain it by showing that it could have been pre-dicted if we had been in possession of the explanatory facts prior to the occurrence. This prediction would have involved a deduction of the ex-planandum-statement from the explanans-statements. On this view we can say that there is a relation of *logical necessity* between the laws and initial conditions on the one hand, and the explanandum on the other.[10]

2.2. *The Modal Conception of Scientific Explanation*

Under the same circumstances we can say, alternatively, that because of the lawful relations between the antecedent conditions and the explanandum-event there is a relation of *nomological necessity* between them. I shall call this view the *modal conception* of scientific explanation.[11] Given the partic-ular set of initial conditions, and the laws of nature, the explanandum-event had to occur. *Nomological necessity*, it might roughly be said, derives from the laws of nature in much the same way that *logical necessity* rests upon the laws of logic. Viewing the matter this way, one can deny that an explana-tion is an argument, but still maintain that the explanation is the sort of thing that shows that the explanandum-event had to occur, given the initial condi-tions. In the absence of knowledge of the explanatory facts, the explanandum-event (the eclipse) was something that might not have occurred for all we would know; given the explanatory facts, it had to occur. The explanation exhibits the nomological necessity of the explanandum-event given the explanatory facts. Although a deductive argument can be constructed (as in the foregoing account) within which a relation of logical entailment obtains, an explanation need not be regarded as such an argument, or any kind of argument at all.

2.3. *The Ontic Conception of Scientific Explanation*

There is still another way to look at such explanations. The term "law" is used sometimes to refer to a scientific statement describing a regularity in nature, and sometimes to refer to the regularity itself. Laws in the former sense *describe* patterns in the physical world; laws in the latter sense *constitute* or *provide* such patterns. Construing the term "law" in either sense, we can say that to relate an explanandum-event to some antecedent conditions by means of laws is to fit the event to be explained into an intelligible pattern. When I call the pattern "intelligible", I do *not* mean to suggest that it possesses any kind of "rational necessity" and I do *not* mean to suggest that such patterns can be known *a priori*. The point is simply that we have formulated the law-statements in terms that we understand, or equivalently, that we have seen and identified the lawful regularity described by the law-statement. In view of the universal character of the laws involved in such explanations, we can also say that, given certain portions of the pattern of events and the lawful relations exhibited by the constituents of the pattern, other portions of the pattern must have certain characteristics. Looking at explanation in this way, we might say that to explain an event is to exhibit it as occupying its (nomologically necessary) place in the intelligible pattern. Because of its emphasis upon existent physical relationships, this view may be called the *ontic conception* of scientific explanation.[12]

2.4. *Laws: Universal vs. Statistical; Causal vs. Non-causal*

These three ways of thinking about scientific explanation may seem more or less equivalent — perhaps with somewhat differing emphases — as long as we are talking about the kind of explanation that involves appeal only to *universal* laws. A striking divergence will appear, however, when we consider explanations that invoke *statistical* laws. In the Laplacian framework, all of the fundamental laws of nature are strictly universal; in twentieth century science, we must at least entertain the possibility that some basic laws of nature are irreducibly statistical.

Before making the transition to consideration of the nature of scientific explanation in contexts where statistical laws must be taken into account, I must acknowledge one factor in the Laplacian conception which did not appear in any of the three accounts. Its neglect would be a glaring omission in any discussion of this sort of explanation. I refer to the relation of *causation*, which certainly played a large role in Laplace's considerations.

It may be tempting at first blush to suppose that the laws of nature are always causal laws, and that explanation in terms of laws is *ipso facto* causal explanation. This view seems implicit in Laplace's discussion, and it has been voiced more or less explicitly by a variety of authors.[13] A moment's reflection reveals, however, that many law-statements do not express causal relations; many lawful regularities in nature are not direct cause-effect relations. Night follows day and day follows night, but day does not cause night and night does not cause day. The ideal gas law

$$PV = nRT$$

relates pressure, volume, and temperature for a given sample of gas, and it tells us how these quantities vary as mathematical functions of one another, but it says nothing whatever about causal relations among them. Kepler's laws of planetary motion describe the orbits of the planets, but they offer no causal account of these motions. Each of these regularities – the alternation of night and day; the quantitative relationship among temperature, pressure, and volume of an ideal gas; and the regular motions of the planets – can be explained causally, but they do not express causal relations and they do not afford causal explanations of the events that are subsumed under them. I shall return to the causal explanation of regularities below.

3. STATISTICAL EXPLANATION (POLLEN)

In 1827, when the botanist Robert Brown first noticed the random dance of microscopic particles of pollen suspended in a fluid, he interpreted it as evidence of their intrinsic vitality – though further observations of other kinds of particles convinced him that this phenomenon had no connection with life. Hé could not have guessed that he had witnessed rather direct visual evidence of the statistical behaviour of molecules of the fluid in which the particles were suspended. That interpretation had to await the publication of one of Einstein's three epoch-making papers of 1905. At that juncture (1905) it was still possible to claim that the apparently random agitations were rigidly determined – just as Laplace had maintained – by the motions of tiny particles which strictly obey Newton's laws of motion. But as the quantum theory developed in the first quarter of the present century, the idea of a deterministic underlying structure became more and more difficult to defend. By now, a large percentage of those who interpret quantum theory maintain that quantum phenomena are fundamentally and irreducibly statistical in character. To consider a well-worn example, the radioactive decay of a uranium nucleus by spontaneous ejection of an alpha-particle is governed entirely by probability. Given two such nuclei, one of which decays while the other does not, the statistical interpretation simply says that there is a certain probability for each of them to decay, and *there is no further factor* which determines that one will decay and the other will not. This is not a matter of human ignorance; it is a fundamental indeterminacy in the world. I *do not* mean to assert dogmatically that this is the correct interpretation; I *do* believe it has to be entertained seriously. Under these circumstances, it seems to me, we need a concept of scientific explanation which can accommodate indeterminacy — a concept of explanation which can handle the irreducibly statistical cases. For if anything is evident as a result of the physics of the last half century, it is that quantum theory has enormous explanatory power.

Let us consider some examples of statistical explanation which are more commonplace. Suppose John Jones has a streptococcus infection from which he recovers quickly after being treated with penicillin.[14] We would naturally explain his quick recovery on the basis of this treatment. However, most, but not all, streptococcus infections respond to penicillin, so we cannot say that he *had* to recover; we can only say that the penicillin treatment rendered his quick recovery highly probable. This explanation falls somewhat short of the Laplacian ideal of showing that the explanandum-event was necessary in the light of the explanatory facts, but it does approximate that ideal in showing that the explanandum-event *was to be expected* with high probability, given the explanatory facts. In admitting such an explanation, we allow for a little looseness or "play" in the system of lawful connections.

Unfortunately, not all cases of explanation obligingly give us high probabilities. If John Smith develops paresis, it is explained by the fact that he contracted syphilis (more precisely, syphilis in the latent stage that has not been treated with penicillin).[15] The incidence of paresis among cases of latent untreated syphilitics is not high; it is less than 50%. This appears to be a case in which an explanation of the explanandum event — the occurrence of paresis — can be given, but it does not render that event highly probable, or even more probable than not. Given an individual with latent untreated syphilis, one should predict that he will *not* develop paresis. What the explanation does afford, however, is a set of conditions which are relevant to the occurrence of paresis, and (at least in our present state of medical knowledge) we can offer no others. We know that no person who does not suffer untreated latent syphilis will contract paresis, but among those who do have untreated latent syphilis, there is no known way of predicting which ones will manifest this form of tertiary syphilis and which will not.

I could continue offering examples of statistical explanations in which the explanandum-event is not highly probable in the light of the explanatory facts – cases in which what is involved in the explanation is quite clearly a suitable assemblage of factors relevant to the occurrence or non-occurrence of the event to be explained. Such assemblages of relevant factors may yield probabilities that are high, middling, or low. The degree of probability is not what counts; the important consideration is to identify the factors that are statistically relevant. If, for example, we want to explain the fact that a particular adolescent became a delinquent, we may find that he comes from a broken home, lives in a neighbourhood with a high delinquency rate, falls within a certain socio-economic class, etc., which makes delinquency highly probable. Another adolescent, from a different home environment, different neighborhood situation, a different socio-economic background, etc., may have a low probability of becoming delinquent – nevertheless he does. The same factors are relevant in the low probability case as in the high probability case, and in my opinion the two explanations are equally adequate. Each appeals to precisely the same probability distribution over the same set of factors relevant to juvenile delinquency.[16]

There is an obvious, but fundamental, point behind these considerations. If, in a well-specified set of circumstances, a given outcome is highly probable, but not necessary, then in some of these cases the improbable will occur. Even if a coin is heavily biased for heads, it will occasionally land tails up. The explanation is exactly the same in both types of cases: this outcome resulted from a toss of a coin with a certain high probability for heads and a correspondingly low probability for tails. If tails does occur, we might remark on its unlikelihood, but this is by way of "gloss". It is not part of the explanation.[17]

In the examples of coin tossing, delinquent behavior, onset of paresis, or recovery from strep infection we believe, quite reasonably, that the cases are not *irreducibly* statistical. We feel very deeply that with additional knowledge of scientific laws, or more specific information about the particular cases, we could say why *this* toss resulted in a tail rather than a head, or why *this* child became delinquent while another in similar circumstances did not. We are apt to feel, consequently, that our explanation is not complete or fully adequate unless we can say why a particular instance constituted an occurrence rather than a non-occurrence of a given outcome. Indeed, this has sometimes been elevated to the status of a *criterion of adequacy* for scientific explanations in general; namely, that one and the same explanation cannot adequately explain either the occurrence or the non-occurrence of a given type of event in the same circumstances. But this is a principle we must relinquish, I believe, if we are to make sense of scientific explanation in a genuinely indeterministic setting. The fact that it is difficult, if not well-nigh psychologically impossible, to give it up is a measure of the degree to which the Laplacian conception of the world permeates our thinking, even if it is just about a half-century out of date.

Let us return to the quantum mechanical example. When an alpha particle forms in a uranium nucleus, it races to and fro inside, repeatedly crashing against the potential barrier that constitutes the wall of the nucleus. In the overwhelming majority of instances it bounces back, but on rare occasions it "tunnels through". All of this can be explained by a quantum mechanical wave function, but that wave function yields only a very low probability (of the order of 10^{-38}) that the alpha particle will escape. Precisely the same wave function explains both the reflections and the penetrations of the barrier; the only difference is that it assigns a high probability to the one result and a low probability to the other.

Some philosophers have maintained that statistical laws give us grounds for prediction, or for assigning fair betting odds, but not explanations. Their reason for denying the possibility of irreducibly statistical explanations is that these do not confer any kind of necessity upon the explanandum-event.[18] We have, I believe, reached the crunch between the Laplacian conception of explanation, which reflects the deterministic world picture of classical physics, and the statistical conception of explanation which is more harmonious with contemporary physics.

When the three general conceptions of scientific explanations were elaborated in the Laplacian context, it will be recalled, they all seemed pretty much equivalent to one another. When we look at them in the indeterministic context, that situation changes remarkably. The first two conceptions, epistemic and modal, involved necessity. The epistemic appealed to the *logical necessity* with which a conclusion follows from the premises of a valid deductive argument. The modal conception invoked the *nomological necessity* with which the explanandum-event is related to the explanatory facts by virtue of universal laws of nature. If either of these formulations is taken as canonical for all acceptable explanations,. then necessity is built into the concept of explanation from the outset. If we accept that conclusion, then indeterminism in the physical world would render scientific explanation impossible or unintelligible. The third conception – the ontic conception – does not have this consequence.

I do not think we are forced to accept any such drastic conclusion. I am convinced that statistical explanations are admissible, and quite possibly indispensable, in contemporary science. In the next sections I shall try to sketch the sense in which statistical explanations, even when the associated probability values are low, provide genuine understanding of the phenomena in question. I shall then say more about the three general conceptions which emerged from the discussion of Laplacian explanation.

4. CAUSALITY IN EXPLANATION

According to the ontic characterization, it will be recalled, an explanation was described as an exhibition of the fact to be explained in its place within the natural patterns of the world. These patterns are based upon the lawful regularities which structure the world. Within the Laplacian framework, these regularities were seen as strict causal laws, but that deterministic feature was not essential to the characterization. It may be, as modern physics suggests, that the laws are statistical at bottom, and the patterns may be probabilistic ones. If this does represent the actual structure of the world, then many (if not all) events will have to be viewed as probabilistic outcomes of stochastic processes. The pattern of the world is then to be viewed as a series of probability relations. It would be a grievous mistake to think that this sort of thing is not a pattern or to suppose that we cannot know or understand it. I should like to attempt to sketch some of the important characteristics of such understanding.

It is customary to make a sharp distinction between causal relations and statistical or probabilistic relations. This dichotomy, it seems to me, should be called into question. Suppose a brick is hurled with great force at a window pane; as the pane shatters we have no doubt that the cause is the impact of the brick. Suppose, instead, that the window is struck by a golf ball traveling with only moderate speed. Under these circumstances, let us say, the window pane will break in 90% of such cases, but not in the other 10%. The motion of the golf ball up to the point of contact with the window is a causal influence, propagated through space, and it produces the effect of shattering in $\frac{9}{10}$ ths of the situations in which it is present. No one would hesitate, I should think, in concluding that it was the impact of the golf ball that caused the breakage in any case in which breakage occurred. The contact of the golf ball with the window obviously has a large influence upon the probability of the window breaking at that particular time; there is nothing like a 90% chance of the window breaking just in the normal course of things, say as a result of internal stresses, the rumble of a passing truck, the explosion of a gas heater in a house three doors down the block, etc. There are, in other words, probabilistic or stochastic influences which — to borrow a phrase from Leibniz — incline but do not necessitate. I see no reason to refrain from calling such influences "causal", even though they are not deterministic. The fact that we may believe that a deterministic explanation *could be given* if more detailed information were available is no objection. The main point remains; we need not commit ourselves to determinism in order to hold that there are causal influences in the world.[19]

Causality has had a bad press in philosophy ever since Hume's devastating critique, first published early in the 18th century. As is well known, Hume analyzed causal relations in terms of spatio-temporal contiguity, temporal priority, and constant conjunction. He was unable to find any "necessary connection" relating causes to effects, or any "hidden power" by which the cause "brings about" the effect.[20]

Hume's classic account of causation is rightly regarded as a landmark in philosophy; it was, I believe, unjustly ignored or unappreciated by writers like Laplace. Nevertheless, it seems to me, Hume did overlook one fundamental aspect of causal processes, namely, that they are capable of *transmitting information*. This feature is crucial, I believe, in assessing the role of causality in scientific explanation. In order to understand this point, it will help to introduce a distinction between *causal processes* and *pseudo-processes*. That this distinction escaped Hume's attention is not surprising, for it has emerged from consideration of Einstein's special theory of relativity (first enunciated in another of his 1905 papers). A basic consequence of that theory is that no *signal* — that is, no process capable of transmitting information — can travel faster than light. For example, radio signals and sound waves are obviously capable of transmitting information; radio waves travel at the speed of light, as do all other types of electromagnetic waves, and sound travels at a much smaller velocity. Certain *pseudo*-processes can, however, travel at arbitrarily high velocities, not limited by the speed of light. If, for example, a rotating spotlight is mounted in the center of a circular room, the spot of light it casts upon the wall can travel at as great a speed as you like, depending upon how fast the light rotates, and how far the walls are from it. There are, to be sure, a number of causal processes involved in this example — the mechanism that rotates the spotlight, the process by which the filament is made to emit light, and the transmission of light *from* the spotlight *to* the wall. All of these processes are subject to the speed limit imposed by nature (as Einstein conceived it) upon all causal processes. The movement of the spot along the wall — though it manifests a high degree of regularity — is *not* subject to such limitation, but it is *incapable* of transmitting information. If, for example, a red filter is placed near the source in the beam of light that travels from the spotlight to the wall, the spot on the wall will be red; the beam of light carries that "mark" or information from the point at which the filter is interposed along the beam to the wall. If, however, a red filter, interposed near the wall, makes the spot on the wall red, the red "mark" will not be carried along by the spot that sweeps around the wall. The spot traveling along the wall does not carry information with it; it constitutes a pseudo-process, not a causal process. This example is analogous to the scanning-pattern on a TV screen. Electrons are shot from a source at the back of the tube toward the screen; the lateral to and fro pattern of electrons impinging on the screen is a pseudo-process. Information is transmitted from the back of the tube to the screen; it is not transmitted along the lines scanned across the screen. It was this ability to transmit information, which distinguishes causal processes from pseudo-processes, that Hume overlooked.[21]

The ideal gas law was cited above as a non-causal law; it does not describe any causal processes. Suppose I have a container of some gas (e.g., helium) with a movable piston. If I compress the gas by moving the piston, without altering the temperature of the gas, we can infer that the pressure will be increased. This increase in pressure can be explained causally on the ground that the molecules, traveling at the same average velocities, will collide with the walls of the container more frequently when the volume is decreased by moving the walls closer together. The quantitative relation between pressure and volume (at constant temperature) is not a causal relation; the motions of individual molecules, obeying mechanical laws and colliding with the walls of the container, are causal processes. This situation is, I believe, rather typical: a non-causal regularity is explained on the basis of underlying causal processes. In a similar fashion, it seems to me, Newton's laws of motion and gravitation, which are causal laws, explain such non-causal regularities as Kepler's laws of planetary motion, Galileo's law of free fall, and the regular ebb and flow of the tides. The regular behavior of the tides had, of course, been known to seafarers for centuries before Newton; indeed, the relationship between the tides and the position and phase of the moon was familiar to mariners prior to Newton, but these mariners did not suppose that they *understood* the rise and fall of the tides on the basis of this lawful relationship.

We can imagine a child on the beach noticing the waves gradually working their way toward the sand-castle he has constructed. Alarmed, he asks why this is happening. A very primitive explanation might consist in informing him of the regular way in which the tides advance and recede. Though citing such a non-causal regularity might temporarily satisfy child-like curiosity, the "explanation" can hardly be considered scientifically adequate – mainly, I am suggesting, because of its lack of reference to any causal influence. The causal explanation of the non-causal regularity does, in contrast, seem to qualify as a reasonable explanation (though not necessarily one that leaves nothing futher to be explained).

Additional examples, of a similar sort, can be taken from biological or social sciences. The efficacy of inoculation against smallpox was known for centuries before the advent of the germ theory of disease, and before anything was known of the mechanism of immunization. The phenomenon of immunity was *understood* only after the underlying causal processes had been discovered. A well-known correlation between slum environment and reading disabilities in young children may exist, and may be said, in a crude way, to explain why a particular child from the slums cannot read. A reasonably adequate *understanding* of this phenomenon emerges only when we have exhibited the causal relations between economic deprivation and failure to learn to read.[22]

5. FUNCTIONAL EXPLANATION (DREAMS)

Early in this paper, I referred to the important role played by functional explanations in a rather broad range of sciences. Freud's dream theory is a particularly striking example of explanations of this type, as are many other explanations in psychoanalytic theory. Such explanations also occur in many other biological and behavioral sciences. Consider a simple biological example. The jackrabbits which inhabit the hot arid regions in the southwestern part of the U.S.A. have extraordinarily large ears. If we ask why they have such large ears, the answer is *not* "the better to hear you with, my dear." Instead, the large ears constitute an effective cooling mechanism. If the body temperature begins to rise, the numerous blood vessels in the ears dilate, and warm blood from the interior of the body circulates through them. The animal seeks out a shady spot, heat is radiated from the ears, and the body temperature is reduced. The jackrabbit has these large ears *because* they constitute an effective mechanism for temperature regulation.

Animals which live in environments like that of the jackrabbit must have some method for dealing with high temperatures in order to survive. There are, of course, many devices which can fulfill this function. Some animals, such as the kangaroo rat, develop nocturnal habits, enabling them to avoid the heat of the day. Other animals, such as humans, perspire. Dogs pant.

From the fact that a given type of animal survives in the desert, we can infer that it must have some way of coping with great heat. Thus, it can be shown deductively — or at least with high inductive probability — that such animals will have *some mechanism or other* which enables them to adapt to the extreme temperatures found in the desert. It does not follow, of course, that the jackrabbit must have developed large radiating ears, or even that it is highly probable that it would do so. Thus, if we want to explain why the jackrabbit has this particular cooling device — as opposed to explaining why it has some mechanism or other which fulfills this function — it seems implausible to claim that we can do so by rendering the presence of large ears either deductively certain or highly probable in view of the available explanatory facts.[23]

The study of social institutions by anthropologists, sociologists, and other behavioral scientists furnishes further examples of functional explanation. According to A. R. Radcliffe-Brown, the distinguished anthropologist who held the Chair of Anthropology at the University of Sydney from 1926 to 1931, social customs can be explained by considering their function or role in society, just as the presence of the heart in mammals is explained on the basis of its function in circulating blood. "Every custom and belief in a primitive society", he writes, "plays some determining part in the social life of the community, just as every organ of a living body plays some part in the general life of the organism."[24] One does not, of course, need to subscribe to Radcliffe-Brown's extreme view that *all* social explanation is functional in order to agree that functional explanations of social phenomena are *sometimes* appropriate.

A classic example of a functional explanation of a social custom is Radcliffe-Brown's study of the joking relationship between a young man and his maternal uncle among the Bathonga in Africa. When the uncle (his mother's brother) is absent, the nephew comes to his hut, carries on a lewd conversation with the uncle's wife, demands food, steals a prized possession of the uncle, and generally deports himself in a disrespectful manner. Such behavior toward any other relative of an older generation, such as a paternal uncle, would be out of the question, and it would be severely censured if it ever did occur. The maternal uncle, on the other hand, is expected to take the nephew's pranks in good humor — without anger, disapproval, or any attempt at retaliation.

Radcliffe-Brown pointed out that among these people kinship relations form a crucial element of the social structure. The disrespectful treatment of certain older relatives by members of the younger generation plays an important role in maintaining the stability of the kinship system. Through detailed analysis, Radcliffe-Brown attempted to show how the joking relationship serves to ease the tensions which naturally arise in kinship systems of the sort found among the Bathonga. Such kinship systems are not entirely different from our own, and the tensions to which he referred are similar to the kinds of in-law problems with which we are familiar. As Radcliffe-Brown explicitly notes, however, in other cultures the function of easing such tensions is fulfilled by other means, such as avoidance of contact with the in-laws.[25] In this case, as in the case of the jackrabbit's ears, a certain function must be fulfilled if a system is to survive. In the case of the jackrabbit, the system is a living organism; in the case of the Bathonga, the system is a social institution. In both of these cases there are functional equivalents — alternative mechanisms which could fulfill the function in question. The same is true of Freud's theory of dreams; many different dreams are capable of fulfilling the same unconscious wish. For this reason, it seems implausible to try to maintain that the existence of one particular mechanism is either certain or highly probable in a given situation.

It has sometimes been claimed that functional explanations are always illegitimate or, at best, incomplete. According to this view, as the biological and behavioral sciences mature and develop, functional explanations will be replaced by explanations of other sorts. Functional explanations, according to this view, may have heuristic value in the early stages of scientific investigation, but they should ultimately be superseded by non-functional explanations. For example, it may be true, *as a matter of fact*, that functional explanations in biology will eventually give way to explanations of a purely physico-chemical sort, but I do not believe that we should commit ourselves to this viewpoint on an *a priori* basis. From a philosophical standpoint, it seems to me, functional explanations may be just as admissible as explanations of any other sort. As long as they play a crucial role in various branches of contemporary science, I do not think they should be ruled out on logical grounds.

Why are functional explanations regarded with widespread suspicion? There seem to be three principal reasons. First, functional explanations have been viewed as teleological and anthropomorphic. This consideration should not deter us, for as I mentioned early in this essay, functional explanations have been purged of teleological elements in such areas as evolutionary biology. This is illustrated by the examples already mentioned. The jackrabbit does not consciously choose big ears to keep his body cool. Likewise, the Bathonga did not consciously choose the joking relationship as a way of easing in-law tensions, and humans do not consciously choose the dreams which are to fulfill their unconscious wishes. Moreover, as is obvious, none of these accounts requires an appeal to the aims of any supernatural agency.

Second, it has sometimes been objected that functional explanations violate a time constraint by explaining the presence of a mechanism in terms of attainment of a *subsequent* goal, rather than on the basis of *preceding* conditions. This objection also is ill-founded. It is because large ears have proved effective *in the past* in controlling body temperature that jackrabbits now have large ears. Even if a particular jackrabbit never required the use of a body-cooling mechanism (if, for example, it were transported to a zoo in a cool locale), the large ears could still be given the same functional explanation (as might be done on a descriptive placard at the zoo). The joking relation among the Bathonga existed when Radcliffe-Brown studied it because (if Radcliffe-Brown's account is correct) it had *previously* succeeded in easing in-law tensions in that society. The occurrence of dreams (if Freud is right) is explained by the *past* success of other dreams in preserving sleep against the disturbance of unconscious unfulfilled wishes.[26]

Third, the most influential theory of scientific explanation during the past three decades has been unable to accommodate functional explanations as such. According to this received view (Hempel [1965]), an explanation is an argument to the effect that the fact to be explained was to be expected, either with deductive certainty or with high inductive probability, on the basis of the explanatory facts. Because typically there are functional equivalents — alternative mechanisms which could fulfill the same function — functional explanations do not, in general, render the explanandum expectable, either with deductive certainty or with high probability. The fact that the received view of scientific explanation cannot account for functional explanations may, however, reflect more adversely upon this philosophical theory of scientific explanation than it does upon functional explanations themselves. If the received view is correct, there are no legitimate functional explanations in science. Some would say, the received view is correct so there are no legitimate functional explanations. But one person's *modus ponens* is another person's *modus tollens*. Others would say, there are legitimate functional explanations in science so the received view is not correct. If we can develop a philosophical theory of scientific explanation which does admit those sorts of functional explanations which do appear to be widely accepted in various branches of science, that fact should, it seems to me, count significantly in favor of the alternative philosophical theory. In the concluding section of this essay I shall try to show that the ontic conception of scientific explanation holds promise of providing just such a theory of scientific explanation.

6. THE THREE CONCEPTIONS REVISITED

In the context of Laplacian determinism, we characterized three general conceptions of scientific explanation: epistemic, modal, and ontic. In that context, the distinctions among the three may have seemed somewhat artificial, but as I remarked, the differences are striking when viewed from the possibly indeterministic standpoint of contemporary science. Some of these features have already been mentioned, but let us bring them together in order to form a coherent overall picture. These considerations are summarized in Table I.

TABLE I
Scientific explanation – three general conceptions

Deterministic	Indeterministic
1. *EPISTEMIC* Logical necessity Argument/deducibility Nomic expectability with certainty 　(vs. unexpected)	High *inductive* probability Inductive support Nomic expectability with high 　probability
2. *MODAL* Nomological necessity Lawful connection with explanatory 　facts Had to happen 　(vs. might or might not have 　happened)	Statistical explanation impossible
3. *ONTIC* Fitting into intelligible pattern Pattern structured by strict causal 　relations 　(vs. haphazard/unrelated to other 　natural occurrences)	Fitting into intelligible pattern Pattern structured by probabilistic 　causal relations Probabilities need not be high

(1) The *epistemic conception*. If determinism is true, then it is always possible in principle to provide a deductive explanation of any event – that

is, to show that it is logically necessary relative to the explanatory facts. If indeterminism is true, then some events are not fully determined by antecedent conditions and laws of nature, so it is not possible, even in principle, to provide deductive explanations of the Laplacian variety. In view of this fact, some proponents of the epistemic approach have loosened the requirements sufficiently to admit that events which are not fully determined may still be explained if their occurrences can be rendered highly probable in terms of statistical laws of nature. In particular, Carl G. Hempel, the leading advocate of the epistemic conception, developed a pattern of explanation, known as the inductive-statistical (I-S) model, which plays precisely this role in the theory of scientific explanation.[27] In making the transition from the deterministic context to the indeterministic context, the fundamental logical relation of deductive entailment is replaced by the relation of high inductive probability. Thus, the requirements for a satisfactory scientific explanation are relaxed in such a way as to allow for the possibility of irreducibly statistical explanations, provided that the event-to-be-explained can be rendered highly probable in view of the explanatory facts. Under these circumstances, we can still say that the event-to-be-explained *is to be expected*, with high probability rather than deductive certainty, in view of the explanatory facts.

(2) The *modal conception*. If indeterminism is true, then there are some events with at least some aspects which are not physically necessitated by antecedent conditions on the basis of laws of nature. With respect to such features of events of this sort, it is simply impossible to show that they *had to occur*, and hence, they defy scientific explanation. I can see no way in which the modal conception can be transformed to enable it to handle explanation in indeterministic contexts. To replace physical necessity with some sort of probability relation would be to relinquish the modal conception, and to move either to the epistemic conception or to the ontic conception.[28] The adherent of the modal conception faces a severe dilemma. Either one makes an a priori commitment to determinism, or one has to deny that quantum mechanical explanations, as they are usually construed, qualify as legitimate scientific explanations. Neither alternative seems tenable.

(3) The *ontic conception*. According to this conception, events are explained by showing how they fit into the physical patterns which are found in the world. In the Laplacian context of classical physics, it appeared that these patterns were strict deterministic patterns; in the light of contemporary physics, it now appears that some, at least, of these patterns are inherently statistical. But this fact poses no obstacle to the construction of scientific explanations. Statistical patterns are bona fide patterns.

Carbon 14 atoms, for example, decay in a statistically regular way, and this regularity provides the basis for the technique of radiocarbon dating, which has proved to be a valuable tool for archaeologists. Other radioactive atoms decay in accordance with different statistical patterns. The half-life of carbon 14 is 5730 years; the half-life of tritium (hydrogen 3) is 12.26 years; the half-life of uranium 238 is 4.5 thousand million years. Among other things, these regularities imply that there is a very high probability that a given tritium atom will decay in a period of 5730 years, there is a 50–50 chance that a given carbon 14 atom will decay in the same period, and there is a very small probability that a given uranium 238 atom will decay in that same period. One important point to be emphasized in this context is that some events fit into statistical patterns with very low probabilities. For example, there is current speculation to the effect that the proton is not absolutely stable, but decays with a half-life of the order of 10^{30} to 10^{32} years. To gain perspective on the time scale involved, it should be recalled that the total age of the universe since the primordial "big bang" is now thought to be about 10^{10} years. Thus, the probability of a given proton decaying within the next year is truly minute. Experiments are now being designed, however, with the aim of detecting proton decays. Even though the probability of any given proton decaying is very small, there is a reasonable chance of detecting such an event if a large enough collection of protons is examined for a few years. Since events fit into statistical patterns with high probabilites in some cases, with middling probabilities in other cases, and with small probabilities in still other cases, the size of the probability of the explanandum-event has no bearing upon the possibility of providing a statistical explanation of it.

The situation regarding statistical explanation can now be summarized. The modal conception does not allow for statistical explanations of particular events.[29] This view seems untenable in the light of the patent explanatory success of contemporary statistical theories in the sciences. The epistemic conception admits statistical explanations of particular events, provided that the associated probabilities are high enough. How high is high enough? That, I believe, is a profoundly embarrassing question.[30] The ontic conception allows statistical explanations of any events which occur within a definite statistical pattern, regardless of the size of the associated probability.

In attempting to make a decision between the epistemic and the ontic conceptions of scientific explanation, the question of whether it is possible to explain events whose occurrences are intrinsically improbable emerges as a crucial one. As a proponent of the ontic conception, I am inclined to give an affirmative answer. There are two main reasons.

First, to maintain that highly probable events can be explained, while improbable ones cannot, seems to involve a strange and arbitrary lack of parity. Consider, for example, a famous genetic experiment conducted by Gregor Mendel. In a particular population of pea plants, he showed that there is a probability of $\frac{3}{4}$ that any given plant will bear red blossoms and a probability of $\frac{1}{4}$ that it will bear white blossoms. Assume that $\frac{3}{4}$ is large enough to qualify as a high probability; if it is not, the example can easily be modified to furnish a higher value. Then, according to the epistemic conception, we can explain the occurrence of a red blossom in that group of plants, but we cannot explain the occurrence of a white blossom. It seems obvious to me, however, that under those circumstances we understand the occurrence of a white blossom just as adequately as we understand the occurrence of a red blossom. The fact that one occurs with a higher probability than the other is beside the point.[31]

Second, as I tried to argue in the preceding section, if functional explanations are to be considered admissible, we will have to allow the possibility of explaining facts which do not have high probabilities. For those who are dubious about the force of the argument from symmetry, given in the preceding paragraph, this argument may be decisive. It is beyond the scope of this paper to attempt to provide a detailed account of functional explanation, but the fact that functional explanations do seem to be considered acceptable in various branches of contemporary science strongly suggests that our conception of scientific explanation ought to be broad enough to accommodate explanations of this sort.[32] Of the three conceptions we have discussed, only the ontic appears to be capable of this.

In my earliest article on scientific explanation, I attempted to develop a theory based upon relations of statistical relevance; this led to the elaboration of the statistical-relevance (S-R) model.[33] Statistical explanations constructed along the lines of this model could accommodate events whose probabilities are low, medium, or high. In more recent writings, I have attempted to supplement the statistical relevance model with considerations of causal relevance.[34] Recalling the claim made in a previous section – that causal relations can also be statistical – we see that the statistical and the causal considerations which have been discussed in this essay can be brought together to form a unified theory of scientific explanation.

Statistical and causal relations constitute the patterns which structure our world – the patterns into which we fit events and facts we wish to explain. Causal processes play an especially important role in this account, for they are the mechanisms which propagate structure and transmit causal influence in this dynamic and changing world. In a straightforward sense, we may say that these processes provide the ties among the various spatio-temporal parts of our universe. We have here, I believe, an answer to Hume's question about the nature of the connections between causes and effects.[35] They are the channels of communication by which the physical world transmits information about its own structure. When we recognize these causal processes, and the role that they play in unifying the patterns into which facts and events fit, then we have gone a long way toward scientific understanding of our world and what goes on within it. The ontic conception thus constitutes a causal conception of scientific explanation which seems to be in harmony with twentieth century science. In recognizing the statistical aspects of causal relations, it provides an appropriate advance beyond the Laplacian ideas which have, until recently, had an almost inestimable influence upon our thought about scientific explanation.

University of Pittsburgh

NOTES

[1] The material in this essay is based upon work supported by the National Science Foundation (U.S.A.) under Grant No. SOC-7809146. I should like to express my gratitude for this support, and to thank colleagues and students, too numerous to cite individually, in Australia and America, for many valuable comments and criticisms. I am also grateful to the University of Melbourne for making possible an extended visit to Australia, where these ideas were discussed under circumstances highly conducive to constructive intellectual work.

[2] Edmund Halley, 'Ode to Isaac Newton', first published as a sort of Foreword to Newton's *Principia*. See Cajori (ed.) [1947], p. xiv.

[3] Lapláce [1951], p. 4.

[4] In Salmon [1978c], I discuss at some length the question of what Laplace's demon must do to achieve scientific understanding. My thesis is that he must go beyond mere ability to infer. This essay is reprinted in Salmon (ed.) [1979a].

[5] This attempt was made by Carl G. Hempel in his [1962a]. My first criticism of this approach was given in my [1965d].

[6] Even if Isaac Newton himself preserved teleological elements in his world-view, they were eliminated by subsequent practitioners of classical mechanics such as Laplace.

[7] Cf. Carl G. Hempel [1962b], esp. pp. 16–19, where Hempel discusses an example from Freud.

[8] This conception of scientific explanation is expressed in Holton and Brush [1973], p. 185.

[9] The term "law" has two quite distinct meanings in the context of discussions of

scientific explanation: on the one hand, it sometimes refers to a regularity that exists in nature; on the other hand, it sometimes refers to a statement that such a regularity obtains. When the distinction between these two meanings is important, and when the context does not make entirely clear which sense is intended, I shall use the phrase "law of nature" to refer to the natural regularity itself, and I shall use such phrases as "law-statement" or "scientific law" to refer to the linguistic entity. In order to qualify as a law-statement, a statement must be true. Statements which have all other characteristics of law-statements, but which may fail to be true, are known as lawlike statements.

10 This is, as a matter of fact, the most widely accepted view of scientific explanation today, at least in contexts where universal laws are available for explanatory purposes. For a clear and thorough discussion of this epistemic approach, both in the deterministic and in the indeterministic contexts, see Carl G. Hempel [1965b]. I attack this "received view" in my [1977g].

11 This conception is expressed by Georg Henrik von Wright [1971], p. 13 and in D. H. Mellor [1976].

12 Although Hempel is to be identified primarily as a proponent of the epistemic conception, he does offer the following characterization of scientific explanation in the concluding paragraph of his major essay: "The central theme of this essay has been, briefly, that all scientific explanation involves, explicitly or by implication, a subsumption of its subject matter under general regularities; that it seeks to provide a systematic understanding of empirical phenomena by showing that they fit into a nomic nexus", Hempel ([1965a]:488). I have expressed a similar idea in my ([1977], p. 162).

13 Most notably, it was suggested in the classic article by Hempel and Oppenheim, 'Studies in the Logic of Explanation' [1948]. Hempel has subsequently rejected this notion; see Aspects, op. cit., p. 352.

14 This example is due to Carl G. Hempel; see Aspects, op. cit., p. 381.

15 This example is due to Michael Scriven ([1959]:478). In the U.S.A., the term "paresis" specifically designates one form of tertiary syphilis.

16 This example, along with its accompanying analysis, was contributed by James G. Greeno in 'Explanation and Information' [1973], pp. 89–91. This article was previously published under the title, "Evaluation of Statistical Hypotheses Using Information Transmitted" [1970].

17 This felicitous way of putting the point is due to Richard C. Jeffrey in 'Statistical Explanation vs. Statistical Inference' [1971], pp. 24–25. This article was previously published in Rescher (ed.) [1969].

18 See, for example, G. H. von Wright, op. cit., p. 13.

19 For a discussion of various theories of probabilistic causality, see my [1980a].

20 It may be that Hume's critique is mainly responsible for the fact that such contemporary authors as Hempel have explicitly denied that scientific explanation must have any causal component. See Hempel, Aspects, p. 352.

21 I have tried to spell out in detail the way in which causal processes transmit or propagate causal influence in my [1977f].

22 In my 'Theoretical Explanation' [1975d], I discuss causal aspects of explanation at some length.

23 We can, of course, explain why this particular jackrabbit has large ears on the ground that it inherited this trait from its parents, both of which had big ears. But if we ask why this trait is present in the species, the answer may be that it originated on the basis of

some sort of chance mutation. The fact that it was perpetuated and propagated is due to natural selection on the basis of its survival value. This point is discussed by Baruch Brody, 'The Reduction of Teleological Sciences' [1975].

[24] A. R. Radcliffe-Brown, *The Andaman Islanders* [1967], p. 229. ·

[25] A. R. Radcliffe-Brown, *Structure and Function in Primitive Society* [1965], Chap. IV.

[26] This point is effectively discussed by Larry Wright in *Teleological Explanations* [1976], Chap. I.

[27] Hempel [1962a]. An improved version is given in Hempel, *Aspects*, pp. 381–403.

[28] D. H. Mellor, "Probable Explanation", op. cit., tries to make this transition, but this appeal to degrees of possibility and necessity strikes me as insufficient for the purpose. It seems to me that the replacement of physical necessity with high *inductive* probability leads to the epistemic conception, while the replacement by high *physical* probability leads to the ontic conception.

[29] In *Aspects*, pp. 380–381, Hempel discusses the deductive-statistical (D-S) model in which *statistical regularities* are explained by deductive subsumption under broader statistical law. Statistical explanations of this sort pose no difficulties for the modal conception, but such explanations cannot explain *occurrences of individual events*.

[30] I have discussed this question at some length in 'A Third Dogma of Empiricism', op. cit., pp. 151–155.

[31] This point is well-argued by Richard C. Jeffrey in 'Statistical Explanation vs. Statistical Inference', op. cit.

[32] Some brief suggestions regarding functional explanation are given in Merrilee H. Salmon and Wesley C. Salmon, 'Alternative Models of Scientific Explanation' [1979c], pp. 70–71.

[33] 'The Status of Prior Probabilities in Statistical Explanation', op. cit., esp. pp. 145–146, and 'Statistical Explanation' in Salmon, *et al.* [1971a].

[34] Mainly, 'Theoretical Explanation', 'A Third Dogma of Empiricism', and 'Why Ask, 'Why?'?' op. cit.

[35] I have elaborated this point in 'An "At-At" Theory of Causal Influence', op. cit.

CLARK GLYMOUR

CAUSAL INFERENCE AND CAUSAL EXPLANATION

Wesley Salmon's account of causal inference and causal explanation is, very briefly, as follows: causality is a feature of processes, a feature they have in virtue of being spatio-temporally connected and of bearing a mark or marks (that is, a property the process acquires in virtue of an interaction with another process) from one space-time region to another, later one, without benefit of subsequent interaction. ("Interaction" is not supposed to be a causal primitive: two processes interact if they occupy a common space-time region — that is, if they intersect — and satisfy a statistical condition.[1]) The explanation of particular occurrences is always causal explanation, but Salmon allows that the concern of science is seldom to explain particular occurrences, and more often is to explain regularities. I am unsure what form Salmon supposes causal explanations, whether of particular occurrences or of patterns, to have, and I am likewise unsure how the emphasis in his more recent essays on causal explanation and the irreducibility of causal relations to statistical relations combines with his earlier emphasis on statistical explanation as a kind of *sui generis* form. I will simply construe his views on these matters in what I take to be both the most clear-cut and the most generous way: there is no such thing as statistical explanation *per se*, merely statistical evidence for (and, perhaps, statistical aspects of) causal explanations. The causal explanation of a particular event consists in the description of any interesting fragment of the causal history leading up to the event. The causal explanation of regularities consists in the description of features of the causal history common to every instance of the regularity. (I admit that Salmon does not say much of this explicitly; but his asides, examples and arguments strongly suggest this reading.) So understood, Salmon's views about explanation are very like those of David Lewis,[2] although they hold quite different conceptions of causal relations. Salmon takes much causal inference to be founded on Reichenbach's common cause principle:

If two or more events of certain types occur at different places but occur at the same time more frequently than is to be expected if they occurred independently, then this apparent coincidence is to be explained in terms of a common causal antecedent.[3]

179

Robert McLaughlin (ed.), What? Where? When? Why?, 179–191.
Copyright © 1982 by D. Reidel Publishing Company.

Such common causal antecedents must be postulated even if they are unobserved or unobservable. Unobserved and unobservable entities and events are warranted for two reasons: in the first place, they permit causal histories to be continuous, and in the second place, they avoid miracles. Here I must quote:

> The fundamental fact to which I wish to call attention is that the value of Avogadro's number ascertained from the analysis of Brownian motion agrees, within the limits of experimental error, with the value obtained by electrolytic measurement. Without a common causal antecedent, such agreement would constitute a remarkable coincidence ... In my opinion, the instrumentalist cannot, with impunity, ignore what must be an amazing correspondence between what happens when one scientist is watching smoke particles dancing in a container of gas while another scientist in a different laboratory is observing the electroplating of silver. Without an underlying causal mechanism – of the sort involved in the postulation of atoms, molecules, and ions – the coincidence would be ... miraculous ... [4]

I have any number of qualms about these doctrines, and it turns out that several of my qualms are interconnected. My reservations are about correctness and completeness both. I begin with a list of theses:

(1) There are no grounds for requiring that causality be characteristic of processes connected in space-time, but there are good grounds for denying that requirement.

(2) Many theoretical entities, states, events, properties, etc., are postulated for reasons having little or nothing to do with a need to sustain the space-time connectedness of causal processes, and some theoretical entities could not posibly play such a role.

(3) The common cause principle is stated in too restricted a form, since the principle is widely and explicitly used in psychology and in the social sciences to infer a common cause for events or states of affairs that are not spatially separated.

(4) The common cause principle is stated in a form too strong to sustain, since in scientific practice we are not always bound to infer a common cause from correlations.

(5) Many scientific explanations derive their explanatory force from considerations that are not causal.

(6) The assessment of competing causal explanations often depends on features that are omitted from the view of explanation as provision of a causal history.

(7) The argument that theoretical claims avoid recourse to miraculous

coincidences is no better than are the common run of numerological arguments.

In most cases these items amount to cavils rather than criticisms, and I surely agree with much of the spirit of what Salmon writes, and some of the letter. Still, the cavils may be of interest.

1

The history of science is filled with disputes over action at a distance; certainly a lot of physicists have supposed that it is not incoherent to postulate causes that act at a distance. Given that history, I think we ought to require a good argument before agreeing that connectedness in space-time is essential to causal processes; Salmon does not really provide us with one, although he does provide us with many examples. But we seem to have at least the outline of a good argument to the contrary: big things are made of little things, and the causal relations between occurrences involving big things are no more than complications of relations among the little things of which the big things are constituted. Causal processes cannot be connected in space and time unless the processes involving the ultimate constituents of matter are so connected. But developments in quantum theory appear to make it very unlikely that the time evolution of systems of fundamental particles is generally connected in the requisite way. Rather than concluding that there is no causality in nature, the reasonable thing, I should think, is to conclude that causality does not require space-time connectedness, and all the more so since action at a distance is no novelty in the history of physics. That leaves us, among explicit theories of causality, those which analyze causal relations as probabilistic relations, and those which analyze causality as a kind of counter-factual dependence. I wholly agree with Salmon in finding probabilistic accounts unbelievable, and I am left, therefore, despite my unease over its metaphysical generosity, with an account of causality as counter-factual dependence such as that advocated by Lewis.[5]

2

Certainly Salmon does not say that the only reason for postulating unobserved or unobservable entities, events, processes, is to maintain space-time connectedness in causal histories, and certainly I do not want to say that there is never such a motive. I want only to claim that such considerations can at best constitute a small fraction of the reasons for transcending what is observed. The occasions I can think of in which a theoretical entity has been postulated because of considerations of space-time connectedness (*not*: the case of postulated entities having a connected history) are fairly rare: nineteenth century field theories of electricity and gravitation, Freud's unconscious mental states, the germ theory of disease, and theories which postulated certain elementary particles, notably Roentgen's hypothesis about x-rays. Doubtless there are others. But there are a great many cases in which connectedness considerations formed no part of the argument for the need to postulate an unobserved entity: Dalton's atoms, Newton's gravitational force, Neptune, Vulcan, black-holes, neutrinos, etc., were not postulated because they saved the connectedness in space-time of causal histories. There are some cases into which connectedness considerations could not possibly enter: absolute space, for one.

3

Philosophical ontologies appropriate to causality are far narrower than scientific ones. Most philosophers allow events as causes, some allow states of affairs as causes; Salmon allows causal processes. Very often temporal conditions are imposed; e.g., the cause and the effect must not be simultaneous. Any reading of scientific journals, and especially of social scientific and psychological journals, will make it pretty obvious that the scientific community is perfectly at home in speaking and writing of causal relations among *properties*, and that scientists sense no incoherence in talking of a thing's exhibiting one property as caused by its exhibiting another property, even though the exhibitions (which might be construed as events) overlap in space and time. Typically, something that happens to a thing, some occurrence involving it, is explained in terms of a more or less enduring feature of the thing. Thus the explanation Archimedes gives for why lead sinks in water is (in effect) that lead is denser than water, and so any particular piece of lead sinks on any particular occasion because that piece of lead weighs more than does an equivalent volume of water. Again, one finds it suggested that particular people do particular things because the people have certain traits: Amos Anonymous scored 112 on an IQ test because his general intelligence is greater than average. More typically, the variations in performance of members of a population, or their variations in some one trait, are explained by their variations in some other trait. Thus the variation of IQ scores in the general population is explained in part by variations in general intelligence in the population.

There is no doubt that this is how scientists talk, and no doubt that if one puts this kind of explanation outside the domain of a philosophical theory of causal explanation then one will have excluded from the compass of one's philosophical theory a great deal of the explaining that goes on in science. The philosophical theories of causality advanced by social scientists are perfectly sensitive to this fact. Thus Herbert Simon's analysis of causality [6] (which I do not endorse) gives an account of causal relations among variables — that is, among what Newton would have called properties admitting intention and remission of degree. If one does not want to speak of causality as the scientists do, then one must either show how to reduce their discourse to talk about causality considered philosophically tolerable, or else one must admit that something distinctly non-causal is involved, so that much of scientific explanation of particular occurrences is, after all, not causal explanation. I don't think that it greatly matters which course one takes; what does matter is that this kind of scientific explanation, and the patterns of inference that go with it, not remain outside of philosophical consideration.

The incongruence between scientific talk of causal relations, on the one hand, and philosophical theory, on the other hand, is especially trying for Salmon's philosophy, because it is exactly in circumstances in which scientists are concerned to explain performances (or traits) by other, generally unobserved traits, that the common cause principle is most explicitly used in contemporary psychology and social science. The standard methodology in contemporary psychometrics is factor analysis, and the very idea behind factor analysis is the principle of the common cause, i.e., correlations among variables are to be explained by a common cause. The variables observed are usually (but not always) performance measures of some sort, and the factors postulated to explain the correlations among such variables are often enduring traits, capacities, etc. Thus correlations among scores on sub-tests of a mental aptitude test battery are typically explained in terms of some unobserved factor or factors which has a causal role in producing individual scores; such factors may be described as general intelligence, or memory, or fluid intelligence, or whatever. Factor analysis is no more than a systematic, and perhaps rather stylized, procedure for applying the principle of the common cause without Salmon's requirement of spatial separation.[7]

4

Moral philosophers are familiar with the notion of a *prima facie* duty: it is a duty, indeed, but a defeasible one that can be overwhelmed by other considerations. Philosophers of science would do well to adopt a like notion, for principles that are stated as rules of scientific inference often have a similar defeasibility. So it is with the common cause. At best, the rule must be that we should infer a common cause from correlations unless we have a good reason not to. Sometimes we do have good reasons not to; Van Fraassen [8] argues that we can characterize statistically dependent events which the quantum theory assures us have no common cause, and I find the argument convincing. But one can think of less esoteric cases. I and probably others would not infer the existence of telepathic powers from experiments that revealed a small but statistically significant correlation between the responses of the supposed ESP sender and receiver and that, to my satisfaction, eliminated all physical mechanisms for transmitting information. Accidents do happen.

5

Undoubtedly a great many scientific explanations are presented as causal explanations; quite plausibly, many explanations that are not presented as causal explanations are nonetheless appropriately understood as species of such − e.g., functional explanations. There remains, however, a considerable bit of science that sounds very much like explaining, and which perhaps has causal implications, but which does not seem to derive its point, its force, or its interest from the fact that it has something to do with causal relations (or their absence). I will give some examples.

Hilbert, Weyl, Droste, Einstein, then many others, sought to account for the phenomena of gravitation and electro-dynamics on the basis of some variational principle. Variational principles are not causal principles at all; they require simply that some functions always have locally extreme values. Now one may think that in particular applications more restricted variational principles are just surrogates for causal claims; thus one may think that in explaining the path of a light ray through a refracting medium by means of Fermat's least time principle, one is simply making oblique reference to causal interactions which everyone interested more or less understands. That is less likely, however, when one is dealing with variational principles for fundamental theories of gravitation and electro-magnetism: one doesn't know what the fundamental causal processes are. Again, one might observe that variational principles issue in a set of differential equations which can be understood as descriptions of causal histories. Two points must qualify this observation. One is that in the context of general relativity, and related theories, it is not at all obvious that the Lagrange equations obtained from a variational principle can always be construed as causal equations like the equations of motion in classical dynamics. (Certainly they cannot always be so construed globally, for there need not exist a global time function; there may be an interesting question as to whether a variational principle for the metric for the equations of motion of general relativity, and related theories, can always locally be construed causally − provided one could be precise about what "construed causally" comes to.) The equations of general relativity, for example, unlike those of classical mechanics, do not generally admit a Cauchy surface.[9] Second, even if it turns out that such variational principles are always equivalent to "causal" differential equations, the demand for an explanation founded on a variational principle is a special sort of demand, not satisfied by just any set of differential equations.

Another kind of example. Einstein, his collaborators and his successors spent considerable effort trying to derive the general relativistic equations of motion from the field equations of general relativity. The basic idea of the derivations is that on the world-line or world-tube of a material particle the stress-energy tensor must be non-vanishing; the field equations imply, furthermore, that the stress-energy tensor everywhere has a vanishing divergence just equal to the divergence of the Einstein tensor. From these conditions one deduces that the world-line must be a geodesic of the affine connection, which is exactly what the equations of motion assert. Is this a causal explanation? Is it an explanation at all? Whatever it is, do we learn anything much about the point, the structure, and the value of such explanations from trying to insinuate causal claims?

A third example. Consider physicists' explanations of why certain events don't occur, or why certain processes cease: why can't the temperature of any solid be-reduced to absolute zero? Because the temperature is proportional to the mean kinetic energy and the ground state for solids has a nonzero vibratory energy. David Lewis[10] says of a similar example (explaining the cessation of stellar collapse in terms of the Pauli exclusion principle) that information about the causal history has indeed been provided, but it is negative information: the cessation of temperature lowering, like the cessation of stellar collapse, had no causes at all, save for whatever causes brought about the initial temperature lowering or the initial stellar collapse, whose occurrence was a precondition for the cessation of the process. Now I don't deny that the scientific answers to the questions − What made the temperature stop dropping? What made the star stop collapsing? − do contain that sort of negative causal information. I don't even deny that the information is interesting. I do think, however, that the point of giving the scientific explanations, the value of the scientific explanations, is scarcely exhausted by the fact that the explanations imply the right thing about the causes, namely that there weren't any. To the contrary, much of the point of such explanations is better caught by the old-fashioned covering-law account of explanation. For what the explanation provides is a general law restricting the possible circumstances in the world, a law which would be violated if the process in question did not cease. Without such laws one could still indicate the causal factors — there were none — but I don't find it at all convincing that one would still have a scientific explanation.

6

Some philosophers — I don't know whether Salmon is among them — would pretty strictly separate the business of giving a philosophical account of what a scientific explanation is from the business of giving a philosophical account of criteria for comparing and assessing the satisfactoriness of explanations. I would not. Explanations are often assessed in virtue of how plausible or likely, in view of other knowledge, their claims may be; such assessments, when varying from context to context, and employing no general principles peculiar to the business of explaining, ought to be ignored when giving a philosophical account of the nature of scientific explanation. But there are also many, many cases where there is little or no external evidence for the claims that go into an explanation, and the claims are argued to be worthy of belief exactly because they explain so well. Because they are, indeed, so explainy. The considerations that are tacitly appealed to in such circumstances had better be revealed by an account of scientific explanation; there is nothing else to reveal them.

Now I maintain that at least some of the features by which we compare competing explanations, and decide between them, are quite left out of sight if we insist that the point of an explanation is to give an informative causal history. Of course, there is no reason why an explanation cannot have this point as well as others. One example, which I have described in more detail elsewhere,[11] concerns the comparison of Ptolemaic and Copernican explanations of planetary regularities. Both theories provide causal mechanisms that generate all of the instances of the regularities, but Copernican theorists claimed that their theory, not the Ptolemaic, provided the better explanation. Ptolemy, although he could well-enough derive all instances of the regularity, did not claim to have explained it at all. Perhaps there is something more to giving an attempt at an explanation than simply giving a hypothetical causal history and relevant statistics. The same point can be illustrated in more detail by returning to the subject of factor analysis.

Psychometricians typically find themselves in the following circumstance. A test battery has been given to a population, and there are correlations among the scores on various sub-tests or items. How are these correlations to be explained? One way is to suppose that there is a causal connection among the scores on test items themselves, possibly through subtle intermediaries, so that scores on the early test items cause scores on the later test items. The causal model then looks something like Figure 1:

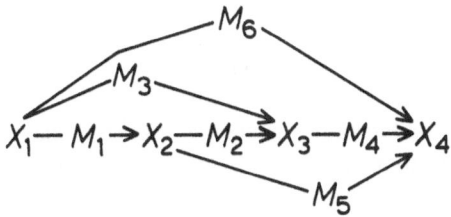

Fig. 1.

Where the Xs are the item scores and the Ms are some intermediary states of the subject. Alternatively, one can explain the correlations by supposing that the item scores have a common cause or causes, perhaps like Figure 2:

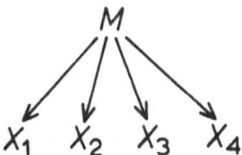

Fig. 2.

Appropriate values for the correlations between the observed variables (the item scores) and the hypothetical factor will then entail the item-item correlations. Unfortunately, if one can obtain the item-item correlations as in Figure 2, one can also obtain them an infinite number of other ways, e.g., Figures 3 or 4.

Fig. 3.

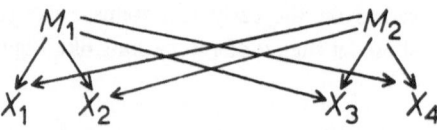

Fig. 4.

Indeed, one has a mathematical guarantee that some factoring can always be done: correlations among n variables can always be generated by n-1 factors, and, of course, sometimes by fewer factors.

Very well, suppose one gives the test battery and computes the correlations among test items, and they can be generated by any of the four models above (as well as others, of course). How does one decide among these hypotheses? Factor analysis itself will not decide, for several reasons. In the first place, factor techniques presuppose that there is a common cause, so that they assume that the correlations are not generated by a mechanism like that of Figure 1. Even excluding Figure 1, factor analysis works in the following way. Partly by guess and partly by system, one obtains a first factor, and its correlations with the observed variables; one then subtracts the covariation due to the common first factor and from the residual covariance among the observed variables obtains a second common factor, and so on. The residual covariances eventually vanish, or become trivially small. However, if one extracts n factors, with given correlations with each of the observed variables, one can always find an infinity of alternative sets of n factors that will also generate the observed correlations. Factor analysts usually represent this point geometrically: let each factor extracted be represented by a unit vector in n-dimensional space, with all such vectors orthogonal. The n factors are then represented by an orthonormal basis in the vector space. Represent each observed variable by a vector satisfying the following conditions: (a) the length of the vector is equal to the proportion of the variance of the variable due to the variance of the several factors, and (b) the cosine of the angle between any two such vectors is equal to their correlation coefficient (so the cosine of the angle between the vector representing an observed variable and the vector representing a hypothetical factor is equal to the hypothetical correlation of the two). Then it is obvious that any alternative set of n unit vectors spanning the space will also provide a set of factors, different from the original, that will generate the observed correlations. Thus even if the factor analysis procedure gives one a single factor having significant correlations with the observed variables, and leaves the residual correlations nearly zero (so that the vectors corresponding to the observables are nearly colinear with a factor vector), a simple change of basis can easily give two (or even more) hypothetical factors that have significant correlations with the observables.

These questions — when is factor analysis appropriate, what number of factors should be extracted, what principles should determine the basis used — are among the most hotly debated in psychological methodology, and have been for some years. It is a pity they are so fully ignored by philosophical methodologists. In any case, I want to point out an old idea, due to Spearman,[12] for settling these questions, an idea which still probably plays a tacit role in the answers some psychometricians settle upon. Suppose that the measured correlations ρ_{ij} in the four observed variables satisfy, or very nearly satisfy, two equations

(1) $\rho_{12}\rho_{34} - \rho_{13}\rho_{24} = 0$
(2) $\rho_{12}\rho_{34} - \rho_{14}\rho_{23} = 0$

Spearman sees this as an indication that the best explanation of the correlations is that of Figure 2. The reason is straightforward, at least at first. Although the causal models associated with all of the other figures can adjust the correlations associated with the various arrows so as to generate the observed correlations and thus also satisfy Equations 1 and 2, only the model of Figure 2 necessitates Equations 1 and 2, whatever the correlations between M and the observed varables may be. Any hypothetical correlations between M and the Xs, consistent with the model of Figure 2, will satisfy Equations 1 and 2, but the other models will satisfy these equations only for very particular values of the correlations between the factors and the observables. Put another way, the point is this: Figure 2 entails equations relating the observed correlation coefficients to combinations of the theoretical factor-item correlation coefficients. If one substitutes for the observed correlation coefficients in Equations 1 or 2 the combination of factor-item correlations with which they are equated by Figure 2, each of Equations 1 and 2 is transformed into a mathematical identity.[13]

There is more than one way to view this preference. Perhaps part of the point of explanations is to eliminate, or deduce, contingency, and that is done by a causal skeleton that transforms observed regularities into instances of necessary truths. Alternatively, one might think, as Skyrms [14] has suggested, that a maximum likelihood principle operates in comparing explanations, and that the example before us is but a special and relatively uncontroversial case of the application of that principle: according to Figure 2, Equations 1 and 2 are as likely as can be – they're certain. Or, one might think, as again Skyrms [15] has elegantly suggested, that the two points of view are ultimately the same. In any case, there is structure here that has to do with how explainy an explanation is, and which ought not to be left aside in accounts of scientific explanation.

7

Many philosophers, including myself, lust for a good argument for scientific realism. Too often we settle for an argument of a kind we would dismiss out of hand if it came from other voices with other aims. I think Salmon's argument about the amazing coincidence between the values of Avogadro's number determined from molecular kinetic theory and from the electrochemical theory of electrolysis is of that kind. I certainly concur with him that the agreement gives us reason to believe the atomic theory; I cannot believe that it does so because, without the atomic theory, the coincidence would be nearly miraculous. The world is full of surprising numerical agreements: there are any number of systems of quasi-physics and plain pseudo-physics founded on little more than the amazing agreement of combinations of physical constants. For years and years a Mr. Norman Bloom, large, intense, unshaven, has presented the faculty of Princeton and other Eastern American Universities with his proofs of the existence of God, based on the fact that the moon's size and distance and the sun's size and distance are precisely such as to permit the moon's disk to totally occlude the sun's disk, and on other amazing coincidences.[16] Mr. Bloom's coincidences are indeed amazing, and if his theories were correct they would be no surprise, but that doesn't make the coincidences good evidence for his theories. Something more is required, some structural connection between the hypotheses and what they purport to explain. The business of philosophers of science ought to be to say what more it is.

I am done arguing for my theses. Some of them are little more than cavils or qualifications or suggestions for further work, and do not really touch in an important way any of the points that I suspect are most dear to Salmon. Only my criticism of action by contact really attacks something cherished. I will not conclude by temporizing, for I do believe that Salmon's analysis of causality is out of step with our physics, but I must also remark that it contains much of what I find most admirably original and refreshing in his writing. One manner of originality is to sustain an idea, or a style, when fashion has long abandoned it, and to bend it, without disloyalty, to the demands of new times. I have been sometimes accused of being the last logical positivist, and I take the accusation as complimentary. Those of us who know and admire Salmon know also that he is really a son of the Enlightenment, a man who belongs in spirit to the age of reason, not to this mad time. And there is an integrity of sensibility and intellect, for Salmon will have none of modal realism, necessary connections, action at a distance, discontinuity or occult powers. He is, I believe, the last Mechanical Philosopher.

University of Pittsburgh

NOTES

1 See Salmon [1978c].
2 Cf. Lewis, 'Causal Explanation', forthcoming.
3 Salmon [1978c], p. 691.
4 *Ibid.*, pp. 698–699.
5 Cf. Lewis [1973].
6 Simon [1977], Sec. 2.
7 See, for example, Cattell [1978].
8 See van Fraassen, *The Scientific Image* [1980].
9 For related points, see Earman [1971].
10 Lewis, 'Causal Explanation', op. cit.
11 In *Theory and Evidence* [1980] and in 'Explanation, Tests, Unity and Necessity', *Noûs*, forthcoming.
12 Spearman [1927].
13 For more detail, see chapter 7 of *Theory and Evidence*, op. cit.
14 Skyrms [1962].
15 Skyrms [1980].
16 For more on Mr Bloom and another numerological eminence, see Gardner [1976].

BAS C. VAN FRAASSEN

RATIONAL BELIEF AND THE COMMON CAUSE PRINCIPLE[1]

During the past decade, Wesley Salmon has successively refined and elaborated Reichenbach's principle of the common cause, as part of a wide-ranging inquiry into statistical inference and explanation. Being as convinced as he of the central importance of the probabilistic concept of common cause, but skeptical of its universal applicability, I have corresponded with Salmon on this subject, and some of that correspondence has made an appearance in both our published writings.[2] In this paper I shall try to state exactly what I think is the significance (especially the *epistemological* significance) of the principle and provide a new (I hope, improved) version of one line of argument in our correspondence. I should emphasize that this has been a very cooperative enterprise: my arguments always began in rather feeble, intuitive form and in answering them, Salmon would gently correct my mistakes, or restate the arguments in a stronger and more precise form than I had managed, and often add striking, concrete illustrations.

1. COMMON CAUSES AND EPISTEMOLOGY

There are coincidences, and some have no explanation, happen for no reason at all. If our universe is indeterministic, that must be so. But if a certain apparent coincidence happens repeatedly, we arrive at a statistical correlation. Then we look for a reason; and Reichenbach's common cause principle says (roughly) that if there is a positive correlation between simultaneous, spatially separate events, then there is a third event in their common past which accounts for their frequent joint occurrence. (I shall give the precise statement below.)

On the face of it, this is an *empirical* statement. It reminds us a bit of certain traditional principles of metaphysics, such as that every event must have a cause. Indeed, Reichenbach proposed it as an acceptable substitute for the deterministic thesis of certain earlier philosophies he criticized. The principle can also be regarded, however, as *methodological*, as marking a criterion of success for science: a scientific theory concerning those correlated events is not complete unless it exhibits, or implies that there is, such a common cause. More weakly, we could take it as a tactical maxim for scientific inquiry: the injunction to proceed by looking for common causes, or to

193

Robert McLaughlin (ed.), What? Where? When? Why?, 193–209.
Copyright © 1982 by D. Reidel Publishing Company.

develop theories in which they are postulated, as a first major line of attack on recalcitrant phenomena.

But there is a further way to look at it, which has to do with rational inference. In epistemology today there is one simple, extreme position, which may be called *extreme Bayesianism*, after the Bayesian statisticians who inspired it. This is the position that a rational person's epistemic state can be represented faithfully and without loss by means of a probability function; that any probability function at all can so represent some rational person; and that rational change of epistemic state consists in conditioning of that personal probability on the total evidence received. I am not sure that any philosopher holds this position in its pure and pristine glory, but it is a useful extreme to consider. There are many challenges to it, and one sort of challenge derives from the old idea of rational non-deductive inference.

If we state Reichenbach's principle as an empirical proposition, there are many probability functions that do not give it a high value. If, secondly, a person manages his garden of beliefs in such a way that, whenever he has a certain degree of belief that two events are positively correlated, he gives at least that degree of belief to the proposition that they have a common cause, then either he gives probability *one* to that empirical proposition (the common cause principle) or else his belief change does not follow the pattern of conditioning on the total evidence.[3] More interesting yet is a combination with the rule of inference to the best explanation: if his theory of explanation implies that the best explanation for a positive correlation is a common cause, and such a common cause is postulated, he will have reason to believe the postulate. As Salmon has rightly emphasized, the principle of the common cause will appear as a powerful argument for scientific realism when it comes in any of these rational inference related forms.[4] It also appears, in that case, as one of the most intuitively appealing examples of a mode of rational belief change that challenges the extreme Bayesian position that coherence of degrees of belief and conditioning are the sole and sufficient hallmarks of rational belief.

That Bayesian position is not one I wish to defend. But I see serious difficulties with the idea that the principle of the common cause should take the strong form in which it rationally compels us to believe in the reality of common causes for all positive correlations.

After stating the principle, and examining a series of examples and arguments, I shall return to these larger questions at the end.

2. THE PRINCIPLE OF THE COMMON CAUSE

To begin, I shall state the principle using the formulae that Reichenbach used (but in standard notation). Then I shall make some remarks on statistical dependence and independence; and based on those, reformulate the principle in equivalent but more perspicuous fashion.

2.1. Statement of the Principle

Two events A and B are called statistically independent if $P(AB) = P(A)P(B)$. When the equality is replaced by the *greater-than* relation we may call them positively correlated. Each of these notions may be relativized to a third event C, using the conditional probability $P(-/C)$. The principle is now that, if

(1) $P(AB) > P(A)P(B)$

then there is an event C such that

(2) $P(AB/C) = P(A/C)P(B/C)$
(3) $P(AB/\bar{C}) = P(A/\bar{C})P(B/\bar{C})$
(4) $P(A/C) > P(A/\bar{C})$
(5) $P(B/C) > P(B/\bar{C})$.

These are the relations stated in their most manageable form, that is, in terms of generic events. We must not ignore the time element, however. The meet AB is an event which happens at a time if and only if both A and B happen at that time. Let us write At for the (individual, non-generic) event which is the occurrence of (generic) event A at t (or if you would rather, for the proposition that A occurs at t). In that case the principle can be stated fully employing the time variables. We can also incorporate clauses such as that C occurs in the intersection of the past cones of the occurrences of A and B.

But these refinements and additions will play little role in what follows. There are two relatively independent questions that may be raised. The first is: is there always an event C at a preceding time such that the above probabilistic relations hold? The second is: if C satisfies the stated conditions, does it follow that C accounts for the correlation (can it reasonably be termed the *cause*)? Extra conditions on C may be contemplated in response to the second question; but I shall restrict myself to the first.

2.2. *Statistical Dependence*

Before looking more closely at what the principle says, let us disentangle some of the statistical relationships that play a role in the discussion. Here are some:

(6) $P(AB) > P(A)P(B)$ *A* and *B* are positively correlated;
(7) $P(A/B) > P(A)$ *A* has a positive dependence on *B*;
(8) $P(A/B) > P(A/\bar{B})$ *B* is positively relevant to *A*;
(9) $P(A/BC) = P(A/C)$ *C* screens off *B* from *A*.

In each case, if the probability function *P* is replaced by the conditional probability $P_X = P(-/X)$, then we use the same terminology, adding the rider "relative to *X*". It will be easy to see how cognate terms, such as "independent", "negatively relevant", and the like are used. Note also that in (6), symmetric terminology ("*A* and *B* are") is appropriate because the relationship is so clearly symmetric in *A* and *B*.

The important point to notice is that there is no need to memorize the terms in (6)–(8), and their cognates, because the ones which are easily confused are actually equivalent (provided all the probabilities involved are well-defined).

To make this precise, let us use the letter **R** to range over what I shall call *positive linear relations* among numbers, defined by the properties:

If $0 \leqslant x$, $y \leqslant 1$ and $0 < b$
then
(I) $x\mathbf{R}y$ iff $bx\mathbf{R}by$
(II) $x\mathbf{R}y$ iff $(b+x)\,\mathbf{R}\,(b+y)$

Note that $=, <, >, \leqslant, \geqslant$ are all positive linear relations.

LEMMA. *If* **R** *is a positive linear relation and* $P(X)$, $P(BX)$, *and* $P(\bar{B}X)$ *are positive, then the following are mutually equivalent*:
(a) $P(AB/X) \mathbf{R} P(A/X)P(B/X)$
(b) $P(A/BX) \mathbf{R} P(A/X)$
(c) $P(A/BX) \mathbf{R} P(A/\bar{B}X)$
The proof is by elementary calculations.

2.3. *Restatement of the Principle*

Using the Lemma, we can now restate the properties of the common cause in Reichenbach's principle in several ways.

(10)   If A and B are positively correlated, then there is an event C such that
    (a) A and B are independent relative to C and also relative to $\bar{C}$,
    (b) C is positively relevant both to A and to B.

(11)   If B is positively relevant to A then there is an event C such that
    (a) C, and also $\bar{C}$, screens off B from A,
    (b)   Both A and B have a positive dependence on C.

3. INDETERMINISTIC STATE TRANSITIONS

My main objection to the common cause principle has been that it demands hidden variables of a deterministic sort, leaving little room for genuine, nontrivial indeterministic theories. The first example I gave along these lines used events which were not logically independent. Salmon objected that Reichenbach's principle is meant to apply to events which are localized, separate, and logically not related. Thus examples of the correlation of events which logically cannot help but be correlated in some way are not relevant.

Since then, I found a lemma in a paper by Suppes and Zanotti which says in effect that the universal validity of the common cause principle requires determinism.[5] The argument is that if a system is not deterministic then we can *construct descriptions* of events which we can *prove* to be correlated, and for which no common cause exists. Hence Salmon's objections seem to apply *mutatis mutandis*; and we must instead look for concrete examples of indeterminism where the common cause principle manifestly requires deterministic underpinnings. We cannot expect a universal refutation, it seems to me: the universe could certainly be so chaotic that any family of physically and logically separate events is totally uncorrelated; there the common cause principle is vacuously true.

Envisage a theory that describes a simple system, defining its possible states, and the properties it may have, with probabilities for states and state-transitions. Suppose we claim for this theory that its probabilities are not "reducible": the system is genuinely indeterministic. Can the common cause principle hold?

To give our imagination a hold, we use familiar examples, suspending our disbelief, in their case, that cited probabilities could not be improved upon. Imagine, then, two chameleons in a sizeable cage; they run and change colour and also, let us say, make themselves rigid. Events that may occur consist in (various parts of) this system having certain properties:

(A)   The first chameleon is red
(B)   The second chameleon is rigid
(C)   The two chameleons are running side-by-side.

There may be further physical characteristics treated in the theory: temperature and illumination in the cage, and so on.

The theory now specifies the possible states of this system, and how the state evolves. This is where theories typically go beyond what we observe. For example, there may be colours which the human eye cannot discern, and the theory may say that in short periods of time, a chameleon will rapidly alternate between these two colours. The theory might say that colour change in a chameleon is always strictly continuous, or observably discontinuous, or discontinuous but not observably so. Also, the theory might specify the possible states partly or wholly in terms of parameters not directly related to observable properties. So it is better to think of the relation between states and events "backwards": the theory specifies a set $h(A)$ of states such that A occurs at time t exactly if the system is in a state belonging to $h(A)$, at that time.[6] For convenience, I shall just use the same letter to denote the event and the corresponding set of states, since we shall discuss only one theory at a time.

3.1. First Example: Equiprobable Outcomes

If the total system is in state x at time t, then there are certain probabilities for its state at time $t + 1$ (counting in seconds from a given origin). In this case, for A and B as above (the first chemeleon being red and the second rigid) we have in the first example I wish to propose:

(12) $P(A \text{ or } B/x) = 1$
         $P(A \text{ or } B/y) = 0$ if $y \neq x$

for a particular state x. Note carefully the time element convention: "$P(A/x)$" is short for "$P(A \text{ at } t + 1/x \text{ at } t)$". The second line implies that if any event Q is a sufficient condition for $(A \text{ or } B)$ then $P(Q) = P(x)P(Q/x)$, which will be helpful.

So if the initial state is x, then a second later the first chameleon is red or the second rigid, and if the initial state is not x, then a second later the first chameleon will not be red and the second not rigid. Let us now add to this

(13) $P(AB/x) = P(\bar{A}B/x) = P(A\bar{B}/x) = \frac{1}{3}$
$P(A/x) = P(B/x) = \frac{2}{3}$

It is clear that x does not screen off A from B because

$$P(AB/x) = \frac{1}{3} < \frac{4}{9} = P(A/x)P(B/x)$$

Yet by the remark at the end of the last paragraph, we can calculate:

(14) $P(AB) = P(x)P(AB/x) = P(x)/3$
$P(A)P(B) = P(x)P(A/x)P(x)P(B/x) = 4\,P(x)^2/9$

the first number being larger than the second if $P(x)$ is less than $\frac{3}{4}$, in which case A and B are positively correlated. That $P(x)$ can indeed be low is easily seen if we let x belong to the set of states not belonging to either A or B, in which case the most that can happen is that the system flips back and forth between x and states in $(A$ or $B)$ so that $P(x) \leqslant \frac{1}{2}$.

The above figures mean that the preceding state is not the common cause of the correlation between A and B. States before the immediately preceding one may be assumed to be "forgotten" (the Markov property), and hence, screened of by that immediately preceding state. Similarly, for any event that happens at the time of (or is part of) the preceding state: it is screened off by that state. Are there any other candidates for a common cause? Not if the description of the process the theory has given so far is complete. So if the common cause principle holds, the description must be incomplete: there must be hidden variables.

The hidden variables needed can apparently be of two sorts: relative to them probabilities of the various events could all be zero or one, or relative to them, the change from one second to the next is a continuous intervening process. The first is determinism; the second I shall examine in the next main section (Section 4)

3.2. Second Example: Perfect Correlation

We start again with the state x but now look at the colours of the two chameleons. Suppose each is capable of four colours: red, green, blue, and yellow. Let the events A_1, A_2, A_3, A_4 be the first chameleon having these colours respectively, and B_1, B_2, B_3, B_4 the second chameleon having them. The theory is as before, but predicts instead, conditional on state x, that the

colour will not be yellow, and that the other colours are equiprobable but
perfectly correlated in the animals:

(15) $P(A_i = B_i/x) = 1$
         $P(A_1/x) = P(A_2/x) = P(A_3/x) = \frac{1}{3}$
         $P(A_iA_j/x) = 0$ if $i \neq j$

for i, j = 1, 2, 3, 4. Let us also assume again that every other state leads to
yellow in the next second, so that $P(Q) = P(Q/x)P(x)$ if Q is a sufficient condition for $(A_1$ $or A_2$ $or A_3)$, or for $(B_1 or B_2$ $or B_3)$. This is only a simplifying
assumption for our calculations; we could instead assume, for example, that
all properties mentioned are independent relative to the other states.

The calculations again show a positive correlation of which the preceding
state is not the common cause.

(16) $P(A_iB_i/x) = \frac{1}{3} > \frac{1}{9} = P(A_i/x)P(B_i/x)$
         for $i = 1, 2, 3$;
         $P(A_iB_i) = P(x)P(A_iB_i/x) = P(x)/3$
         $P(A_i)P(B_i) = P(x)P(A_i/x)P(x)P(B_i/x) = P(x)^2/9$

and so A_i and B_i are positively correlated if $P(x)$ is not zero; but the preceding
state does not screen off A_i from B_i.

The remarks about the candidacy of other events for the role of common
cause, which I made at the end of the preceding subsection, apply again.

3.3. *Formal Structure of the Examples*

In my first example I was careful to use as A and B events which are logically
independent, and even physically independent according to the theory since
$AB, A\overline{B}, \overline{A}B$ and $\overline{A}\overline{B}$ can all occur. In the second I was not careful about that:
a chameleon cannot be red and green all over. However, this is an inessential
feature of the example, since it would have sufficed to take different properties such that

$$P(A_1 \text{ } or A_2 \text{ } or A_3/x) = 1$$
$$P(A_iA_j/x) = 0 \quad \text{if } i \neq j$$

which requires no logical interdependency at all.

It may now be noticed that the two examples are formally the same
(Figure 1):

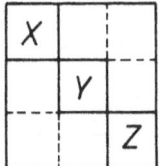

Fig. 1.

First example:

$$A = (X \text{ or } Y)$$
$$B = (Y \text{ or } Z)$$

Second example:

$$X = A_1 \, B_1$$
$$Y = A_2 \, B_2$$
$$Z = A_3 \, B_3$$

Probabilities in both cases:

$$P(X/x) = P(Y/x) = P(Z/x) = \tfrac{1}{3}$$
$$P(X \text{ or } Y \text{ or } Z/y) = 0 \quad \text{if } y \neq x$$

Again, X, Y and Z are chosen disjoint but not exhaustive, but it would suffice to give their intersections probability zero conditional on x, so that the corresponding events could be logically and even physically independent, while not statistically independent. We see therefore that the basic recalcitrant examples can be produced quite easily.

As I noted at the end of Section 3.1, the advocate of the common cause principle need not feel totally stumped at this point. For he can propose that the situation is indeterministic, but that there are hidden variables describing a *continuous*, if indeterministic, process which must underlie the appearances described. Personally I find such a defense as unacceptable as determinism itself, because it also seems to me to run counter to the requirements modern physics accepts as binding on its theories. In any case, I shall devote the next section to the question of how successful such a defense can be. Finally, an advocate of the principle could deny that such examples as I give should be admitted as genuine possibilities. They are, after all, examples of apparent action at a distance without intervening physical mechanism of action-by-contact. They are what Reichenbach would call "causal anomalies". In the

section after next I shall argue that modern physics makes it more reasonable
to accept such possibilities than to deny them. In other words, the principle
appears not to be a regulative ideal of modern scientific practice.

4. INDETERMINISM AND CONTINUITY

As was pointed out to me by Salmon, and as I noted above, the common
cause principle may be satisfied by the introduction of hidden variables,
eliminating *either* the indeterminism *or* the discontinuity. At first I argued
that in all but a trivial class of cases, indeterminism requires the existence of
some discontinuity. Salmon showed me the fallacies of these arguments.
Salmon himself suggested in discussion the example of a lead bullet split
into two parts when it hits a steel knife edge, the paths of the fragments
being correlated, and finally hitting a screen (see figure 2).

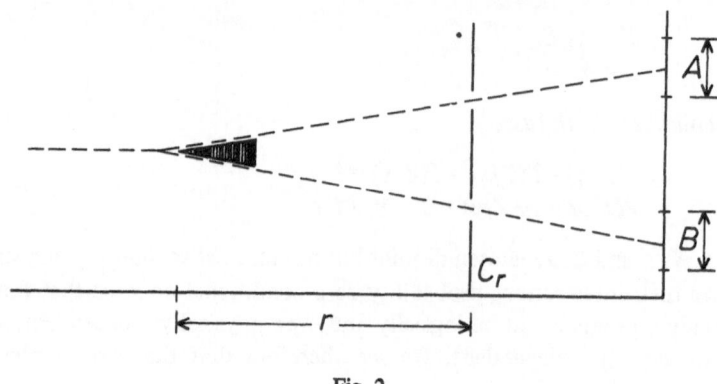

Fig. 2.

Suppose that $P(A \equiv B) = 1, P(A) = \frac{1}{4}$, for example. No finer prediction might
be possible if the scattering is indeterministic. Yet we can see a common
cause: imagine a screen with certain holes in it positioned at a distance r from
the knife edge in such a way that the two fragments go through the holes if
and only if they are to hit A and B. When that happens (call the event C_r)
then both A and B happen, and otherwise neither does. Relative to C_r, they
are mutually independent (all factors equally 1) and relative to $\overline{C}_r$ also (all
zeroes).

   At distance r we also witness a correlated pair of events: each bullet fragment passes through the relevant imaginary hole in the imaginary screen, if
and only if the other does. However, that correlation has a similar explanation

in terms of the events happening meanwhile at distance $r/2$ from the knife edge. So each observable correlation at any distance beyond the edge has an explanation in terms of preceding events at some smaller distance from the edge.

Let us momentarily call the regions (imaginary holes) that the upper and lower fragments must pass through at distance r, in order to hit A and B respectively, by the names $C(A, r)$ and $C(B, r)$. Consider now the very large event $C(A)$ which is the upper fragment passing through all the regions $C(A, r)$; and the corresponding large event $C(B)$ thus related to the lower fragment. We note that the time taken by either (if they happen at all) consists in the interval of all moments after the impact on the edge until the real screen is hit. The events $C(A)$ and $C(B)$ are once more perfectly correlated, and what happens at the knife edge does not satisfy the conditions defining a common cause.

Salmon objected to this that the principle is to be applied only to approximately instantaneous events.

I do not believe that he was happy with this move. It seems to rule out many of our ordinary examples of correlations that are explained by common causes. But I think that we can find a more difficult example.

Let us shift our attention now from chameleons and bullets to roulette balls. I propose a simplified roulette table, a circle divided by three radii into equal red, black and orange parts. There is a concentric circle that cuts each radius at the halfway point; *zeroes* are written in each coloured part inside the inner circle and *ones* outside it. We have two roulette balls, one white and one yellow; and events are of the sort :

At time t, the white ball is in area *red-zero*.

The states of the two balls are perfectly coordinated, as they roll interminably around the table. The yellow ball is always inside the inner circle, and the white ball always outside it; at any time they are always on the same colour (or located on the same dividing radius). Thus the line that joins them is a moving radius of the table. We postulate that its movement is continuous, its velocity unpredictable, and in principle unlimited.

What I mean is this: for each finite amount of time d, we have a probability distribution P_d on the possible positions of the balls at $t + d$ given their position at t. This distribution is such that no matter how small d is, if R is a finite radial triangle on the table, then the probability that the balls' centers will lie in R at $t + d$ is not zero (for any position they may have at t). Thus if

v is an average velocity great enough for the balls to reach R within amount of time d, then they could achieve that average velocity in $(t, t + d)$.

There is no implication of infinite velocity. Suppose that within one microsecond, the balls are found to have moved around the table to immediately behind where they were. Well, a certain finite, but great acceleration will have sufficed to bring them there. If we assume that their acceleration continuously increases, for example, we will have a case in point.

It will be clear that for any time interval $(t, t + d)$ we now have an example that is just like that of our chameleons; and no information about where the balls are at $t + d/2$ (unlike in the case of the bullet fragments) eliminates the uncaused correlation.

5. INTRODUCTION OF INTERACTIVE FORKS

In his presidential address to the Pacific Division of the American Philosophical Association in 1978, Salmon introduced a major innovation into the theory of the common cause (see note 2). With reference to Compton scattering, Salmon replaced clause (2) in the principle with a disjunction: either (2) is the case, or else

$$(2^*) \qquad P(AB/C) > P(A/C)P(B/C)$$

The other clauses stand as before. Thus my second example in Section 3, and its variants that may arise in such a Lucretian world as described in the last section, are accommodated because the preceding states were disqualified as common causes because (2^*) rather than (2) held. I do not mean that the change was made simply because of my examples: Salmon found it easy to show one would arrive at this amendment on other grounds, and that it had independent justification.

Certainly, in Compton scattering the impact of the original photon, and in the half-bullets example the knife edge impact, are the events intuitively cited as the causes. Following Reichenbach, Salmon calls the triple C, A, B a *conjunctive fork* if clauses (2)–(5) hold, and he introduced the name *interactive fork* for the case in which (2^*), (3)–(5) hold. The new intuitive restatement of the principle now possible is:

(17)   If A and B are positively correlated then there is an event C such that C, A, B is either a conjunctive fork or an interactive fork.

There are, of course, qualifications omitted from (17); the correlation is of simultaneous occurrences of A and B, and for each such pair of simultaneous occurrences there must be a preceding occurrence of C. We may need to impose further qualifications on the relationship of C to A and B, before C can rightly be called the common cause when all stated conditions are met, but I shall continue to leave that problems aside.

I have two objections to the new version of the principle. The first is that in the first example I gave in Section 3, $P(AB/x) < P(A/x)P(B/x)$. So there the preceding state does not form either a conjunctive or an interactive fork with the correlated simultaneous events at the next considered moment.

This seems to me an important point. In that example, there were two sorts of states: the special one x, and the others. Given x, A and B are *negatively* correlated. For given x, it follows simply that the system goes into one of the three sorts of states in $(A \text{ or } B)$. Given any other sort of state y, the system goes next into x, which is in $(not\ A\ and\ not\ B)$, so given those other states, A and B are independent $(P(AB/y) = P(A/y)P(B/y) = 0)$. And yet, if we look at the history as a whole, we find that A and B are positively correlated:

$$P(AB) = P(x)P(AB/x) = P(x)/3$$
$$P(A) = P(x)P(A/x) = 2P(x)/3$$
$$P(B) = P(x)P(B/x) = 2P(x)/3$$

Hence, if $P(x) = \frac{1}{2}$ then:

$$P(AB) = \frac{1}{6}$$
$$P(A) = P(B) = \frac{1}{3}$$
$$P(A)P(B) = \frac{1}{9}$$

This is one of these cases of Simpson's paradox, so ably exploited by Nancy Cartwright elsewhere, in which overall correlations are reversed when we look at the infrastructure of the example.

The second objection is that any interactive fork gives rise to a new correlation that falls under the principle; this new correlation requires explanation, in terms of a further fork; and so on, until we reach a *conjunctive* fork. When we say that two events A and B are positively correlated, we tacitly refer them to a certain reference class, or (to use the terminology of a different approach to probability) to our background information. For example, we consider the event of x becoming red, given (if only tacitly) that x is a chameleon of mature age and healthy constitution. What all can go into this given? That depends on two things: what information we have (the question of why A and B are correlated does not arise if the correlation is deduced from suppositions not presently accepted) and on our interests (we don't let the *given* include our knowledge that the little beast has in fact turned scarlet). But I don't think there are other limitations. Suppose now that I remark:

$$P(A/B) > P(A)$$

and you explain by saying, in part,

$$P(A/BC) > P(A/C)$$

and noting that C, A, B form an interactive fork. Then I can add C to my given, and ask why

$$P_C(A/B) > P_C(A)$$

where P_C is the function

$$P_C(X) = P(X/C), \qquad P_C(X/Y) = P(X/YC).$$

And if you accept the common cause principle, then this seems to be a serious and legitimate question: why are A and B positively correlated still, now that we have new information C? Why does the interaction produce a correlation?

Perhaps we can sometimes find an infinite sequence C_1, C_2, ... such that C_{i+1} forms an interactive fork with A and B relative to C_i. In that case each new why-question arising this way does eventually get answered, even though at every stage in the dialogue some remain unanswered. Thus the principle generates a sequence of explanation requests, which terminates only if a conjunctive fork is reached.

6. SHOULD THE EXAMPLES BE TAKEN SERIOUSLY?

Admittedly, running chameleons, coordinated bullet fragments, and roulette balls are fanciful. The most fanciful aspect is perhaps that they are cases of perfect correlation within a world of chance. Their point would not be affected, if we made the correlations less than perfect, though their complexity would be much increased. (I assume this could be done.) But fanciful they would remain, because it would still be as if each member of the pair "knew" what the other was doing and "followed suit". Many successful cases of scientific theorizing have consisted exactly in the reduction of such apparent telepathy to intervening processes acting by contact. (Indeed, one general opinion of telepathy proper is that a successful scientific account of it will do exactly that.)

The reason I feel that we should take such examples seriously is that the development of quantum mechanics shows, it seems to me, that scientific theorizing is not constrained by principles unconditionally demanding their elimination. I do not mean that the sorts of examples I have actually given appear in structurally identical form in quantum mechanics. On the contrary, I have made a special point of giving examples that could conceivably be realized in a Lucretian world which is familiar except in its apparent causal anomalies. But in quantum mechanics we can point to very similar cases, where the development of the theory rejected a demand for mechanisms eliminating such anomalies.

One case is Pauli's exclusion principle. Why can't two electrons in an atom be simultaneously in the same state? How does each "know" to stay out of the states the others are in? Postulation of special "Pauli forces" does not work; other proferred explanations are frankly metaphysical.[6] There are also the examples centering on the Einstein-Podolsky-Rosen paradox. In its most striking form, an atom cascading down from an excited state emits pairs of photons in opposite directions. If we place polarization filters in their way, we cannot predict whether the photons will pass the filters or not, but we can predict that if the filters are oriented at right angles to each other, then either both photons pass or neither does. Among the quantum-mechanical states we cannot find a common cause for this correlation, at least as Reichenbach defines it. "Local" hidden variable theories which can be constructed to explain these phenomena without action at a distance, must all contradict some of the empirical predictions of quantum mechanics itself.[7]

7. THE STATUS OF THE COMMON CAUSE PRINCIPLE

Science is always a search for hidden variables. The models scientists construct have, in all cases but the very simplest, parameters in addition to those they use in their experimental reports describing the phenomena. This is not peculiar to modern science; not only did the atomists set an easily recognizable example, but the Aristotelian use of a plethora of properties and powers, their introduction of properties which are occult (in the technical sense: not reducible to the *compositio*) are another example. Tackling a problem by introducing hidden variables is so pervasive in science that to point it out may be banal. The salutary effect it has, when successful, is *to identify the small but chaotic phenomena we observe with fragments of a large but simple structure* — size being measured here by the number of parameters and elements. Embedding loosely related observable properties in an algebra of occult attributes, or of the kinematic behaviour of a few bodies in a system of Newtonian mechanics with forces and masses and absolute space added to the kinematic quantities, are but two examples of this same procedure.

The enterprise needs a methodology, the grand strategy needs principles of tactics. In the case of the construction of statistical models the principle of the common cause furnishes an important tactical maxim. Faced with correlation of nicotine stains and lung cancer, simultaneously observed in the same patients, we naturally turn to a common cause that will explain both. If we find a conjunctive fork, we are satisfied; there is nothing left to do, the significant phenomenon has found its rightful place in an orderly structure of events.

It was therefore an unprecedented step for the Copenhagen school to take the position that *yes*, there were positive statistical correlations in nature that quantum theory could not eliminate by exhibiting preceding conditions relative to which those correlations disappear, but *no*, the theory was not incomplete.

That this was the Copenhagen position was not appreciated at first. To begin, it was considered outrageous enough simply to insist that the probabilities, whatever they were, could not be "reduced" — that is, could not be regarded as measures of our ignorance concerning underlying deterministic processes.

And so it was very reasonable for Reichenbach, who was as far as I know the first to codify that pervasive feature of statistical model building which we have discussed, to believe that the principle of the common cause is consistent with a belief that the world is not deterministic.

But it seems to me that the two cannot really go together. I do not mean that every indeterministic theory must have the radical features of quantum theory. In that case, it can be *proved*, under certain general (though not entirely uncontroversial) assumptions, that if the theory is true, then there can be no hidden variables allowing a "reduction" of the probabilities. Many probabilistic theories, surely, may be true and yet *compatible* with the thesis that there is a more complete true theory on the same subject in which there is no element of chance. But a belief in the principle of the common cause implies a belief that there is in the relevant cases not merely a compatibility (so that deterministic hidden variables could be introduced into models for the theory) but that all those hidden events which are the common causes, are real, and therefore, that the world is really deterministic.

So, to return to the questions of epistemology, with which I began, I cannot regard the principle of the common cause as a principle of rational belief or inference. It is in my opinion, instead, *a tactical maxim* of scientific inquiry and theory construction. There is no belief, in the case of tactics, that they must be successful. Whether there can be epistemological principles that rationally compel beliefs going logically beyond what we accept as data, is a large question. In the case of the present principle, I am inclined to say that its acceptance does not make one irrational; but its rejection is rationally warranted as well.

University of Toronto

NOTES

[1] Research for this paper was partly supported by a National Science Foundation grant, and it was written while I was released from teaching duties by the University of Toronto Connaught Senior Fellowship. Besides my main, and very large debt to Wesley Salmon, I benefitted greatly from correspondence and conversations with Nancy Cartwright, Ellis Crasnow, Joe Hellige, Jonathan Katz, Edwin Levy, Ben Rogers, and Ron Webster.

[2] See my [1977] and Salmon [1978c].

[3] Both this sentence and the next can be falsified by a trivializing interpretation of conditioning.

[4] See W. Salmon [1975d], and Chapter Two of my *The Scientific Image* (Oxford: Oxford University Press, 1980).

[5] Van Fraassen [1977]; Suppes and Zanotti [1976].

[6] See my [1972] and my summary there of Salmon's comments on that paper at the American Philosophical Association, December 1969, where we began our discussions on causality.

[7] See my 'The Einstein-Podolsky-Rosen Paradox' [1974] for discussion of the original thought experiment and Compton scattering. See J. Bub [1974] Chap. VI, for an exposition of the Bell—Wigner locality argument, followed by a controversial argument concerning its significance. See further my "The Charybdis of Realism: Epistemological Implications of Bell's Inequality", to appear in *Synthese* 52, forthcoming.

J. C. FORGE

PHYSICAL EXPLANATION: WITH REFERENCE TO THE THEORIES OF SCIENTIFIC EXPLANATION OF HEMPEL AND SALMON†

INTRODUCTION

A significant feature of Hempel's writings on scientific explanation, and one which is shared by many who adopt his views as a point of departure, is the lack of examples from modern science. The example which receives most attention in Hempel's long essay on explanation ([1965b], pp. 335–338) concerns the expansion of soap bubbles around tumblers on a draining board. Although the explanation given for this phenomenon is a scientific explanation, it is certainly not an instance which scientists would take to be representative of explanations in science. The use of examples in which familiar states of affairs are explained by elementary scientific laws is quite acceptable if this serves to illustrate the structure of scientific explanation. However, theories of scientific explanation are supposed to cover explanations in science; if instances of explanation in science do not conform to a theory, then that theory is unacceptable even if it can deal with scientific explanations of familiar events.

The purpose of this paper is to examine the consequences of Hempel's theory of explanation for physical science, and to compare these with those of the most recent and interesting alternative theory of explanation, which is due to Salmon. Physical science is certainly one of the most important branches of natural science; it could perhaps be argued that physical science is the paradigm of science. Therefore, an adequate theory of scientific explanation should at least provide a satisfactory account of physical explanation. An examination of the consequences of the rival theories of explanation of Salmon and Hempel for explanation in physical science is an interesting way to test these theories. I shall restrict discussion to explanation of individual events by deterministic laws because this is sufficient to illustrate the strengths and weaknesses of the theories.

1. THE THEORIES OF SCIENTIFIC EXPLANATION OF SALMON AND HEMPEL

Hempel's theory of scientific explanation has been developed in a series of

211

Robert McLaughlin (ed.), What? Where? When? Why?, 211–229.
Copyright © 1982 by D. Reidel Publishing Company.

publications. The first of these [1948] was a joint publication with Oppenheim in which he put forward what has come to be known as the deductive-nomological or D-N model of scientific explanation. In later work (e.g., [1962a] and [1965b]), Hempel extended his theory of explanation to cover explanations given in terms of statistical laws by introducing further models. These models are familiar and I shall not describe them again here (see Hempel [1965b]). For the purposes of this discussion, I shall assume that Hempel's theory of explanation can be stated as follows:

> Scientific explanations of phenomena are arguments which lead us to expect that the phenomena will occur. The logical structures of such arguments are specified by the D-N, I-S, etc., models.

Hempel's theory of scientific explanation is clear, precise and interesting, and so it is not surprising that it has provoked a great deal of discussion and criticism. In particular, a number of criticisms have been levelled at Hempel's claim that explanations are arguments. Salmon, who has been one of the most persistent critics of Hempel's theory, refers to this claim as a third dogma of empiricism (Salmon [1977g]). One source of difficulties for the Hempel theory is, as Salmon points out, that it does not allow for the explanation of events which have low probability. Unless a set of premises supports a conclusion which describes an event to a degree which is sufficient for us to have good grounds for expecting the event to occur, there can be no scientific explanation on the Hempel theory. However, it does appear that it is possible to explain events which have very low probability, once they are known to have occurred. For example, the probability of the spontaneous emission of an alpha particle from a radioactive substance may be extremely low (Salmon [1977g], p. 152). Nevertheless, given that this has occurred, then it may be explained in terms of the particle 'tunnelling' through the potential barrier of the atom. Tunnelling is extremely unlikely; but it is the only way in which spontaneous emission of alpha particles is allowed according to quantum theory, and hence reference to tunnelling explains the event. Thus, it appears that the Hempel theory excludes certain instances of explanation in science.

In addition to excluding *bona fide* explanations, Hempel's theory is susceptible to counter-examples which show that it is too inclusive. For example, given the angle of refraction r of a light ray emerging from a block of glass, the refractive index μ of the glass and Snell's law

$$\frac{\mathrm{Sin}\ i}{\mathrm{Sin}\ r} = \mu \qquad\qquad (1)$$

it is possible to infer the angle of incidence of the light ray impinging on the glass block. The derivation of the angle of incidence satisfies the requirements of the D-N model, and so it qualifies as a scientific explanation on the Hempel theory. Now it is plausible to maintain that the angle of incidence explains the angle of refraction, although we may wish to refer to the position of the light source to give a more satisfactory explanation. But it does not appear to be plausible to hold that the angle of refraction explains the angle of incidence. We certainly do not want to say that the angle of refraction explains the position of the light source relative to the glass block. The reason why we may be reluctant to admit that the angle of refraction explains the angle of incidence is that this is explanation of an event E_1 in terms of an event E_2 which occurred later than E_1. It is difficult to see how the occurrence of an event at time t_0 can be explained by reference to an event which has not occurred by t_0. If it is accepted that scientific explanations exhibit temporal asymmetry in that reference to later events does not explain earlier events, then examples such as that just mentioned show that the D-N theory is too inclusive (Salmon [1977g], pp. 158–162).

Salmon has proposed a theory of scientific explanation which is free from these difficulties. This theory has been developed from the statistical relevance or S-R model of explanation (see his [1970b]). The fundamental notion of the S-R model is the relation of *statistical relevance*. Statistical relevance is a relation which holds between properties or attributes which is such that a property C is statistically relevant to a property B if and only if

$$P(A \cdot C, B) \neq P(A, B) \tag{2}$$

where $P(A, B)$ is the probability that an individual which belongs to the reference class A has the attribute B and $P(A \cdot C, B)$ is the probability that an individual which belongs to a subclass $A \cdot C$ of the reference class A determined by the property C has the attribute B. Salmon summarises the model of explanation based on this notion as follows:

According to an alternative account of statistical explanation [the S-R model] . . . , an explanation consists not in an argument but in an assemblage of relevant considerations. On this model, high probability is not the desideratum; rather, the amount of relevant information is what counts. According to the S-R model, a statistical explanation consists of a probability distribution over a homogeneous partition of an initial reference class.

. . . The goodness, or epistemic value, of such an explanation is measured by the gain in information provided by the probability distribution over the partition. If one and the same probability distribution over a given partition of a reference class provides the explanations of two separate events, one with a high probability and one with a low probability, the two explanations are equally valuable ([1977g], p. 154).

Thus, the question "Why is this x, which is a member of reference class A, also a member of the attribute class B?" is answered by citing a series of probability statements such as

$$P(A \cdot C_1, B) = P_1$$
$$P(A \cdot C_2, B) = P_2$$

.

.

.

$$P(A \cdot C_n, B) = P_n \tag{3}$$

which represent an assemblage of statistically relevant considerations; and by noting which of the subclasses $A \cdot C_i$ contain x. It is not necessary that the P_i are high, and so it is possible to explain events which have low probability.

Salmon is no longer satisfied that an assemblage represented by (3) constitutes a scientific explanation ([1978c], p. 699). In his most recent writings on scientific explanation, he has adopted the view that scientific explanations are causal explanations:

To give a scientific explanation is to show how events and statistical regularities fit into the causal network of the world ([1977g], p. 162).

However, this does not mean that the S-R model is entirely abandoned. The assembly of factors statistically relevant to the occurrence of an event which is to be explained is now the first stage in a scientific explanation. What is necessary, in addition to this, is a causal account of the statistical relevance relations ([1978c], p. 699). It is in this way that the event is fitted into the causal network of the world.

If Salmon is to give a theory of scientific explanation which involves the claim that certain statistical relevance relations are to be given causal explanations, then he is obliged to give some analysis of the concept of cause, because it is not immediately obvious how statistical relations are causally explained. Salmon provides an analysis of cause in his [1978c].

According to Salmon ([1978c], p. 696), it is causal processes, and not causal interactions, which are basic. The transmission of light is a causal process, whereas the absorption or emission of a photon by an atom is a causal interaction. Not all processes are causal processes, some only seem to be causal processes. Reichenbach's mark criterion is used to distinguish causal processes from pseudo-processes ([1978c], p. 690); a causal process can transmit a mark whereas a pseudo-process cannot. Thus, if a car is literally marked by being dented, then this mark is transmitted to the shadow thrown by the car on the roadside. But it is not possible to mark the shadow in the same way; distortions of the shadow as it passes over posts and stones by the roadside are not transmitted to the car. Salmon does not wish to attribute any mysterious power to a process which transmits a mark. He explains how a mark gets from one place to another by employing Russell's "at-at" theory of motion (Salmon [1977f]). A mark gets from A to B by being *at* the intervening points in space *at* the appropriate instants in time.

This analysis of cause, in which a distinction between causal process and causal interaction is made and in which causal processes are distinguished by the mark criterion, enables Salmon to give causal explanations of statistical relevance relations. For example ([1978c], pp. 688–689), the distance of an individual from the centre of an atomic blast is statistically relevant to that person's contracting leukaemia. This is explained by reference to a causal process and causal interactions. The causal process is the transmission of radiation from the explosion to the person, the causal interactions are the emission and absorption of this radiation at either end of the process. The absorption of radiation by the victim may lead to leukaemia.

We have seen that Salmon's theory allows for the explanation of events which have low probability, and so his theory is free from one of the problems which confront the Hempel theory. By requiring that scientific explanations are causal explanations, Salmon also avoids counter-examples of the kind mentioned above. It is not possible to give a causal explanation of the angle of incidence of a light ray impinging on a glass block, nor of the position of the light source, in terms of the angle of refraction, and hence a purported explanation of the angle of incidence by reference to the angle of refraction is not allowed as a scientific explanation on Salmon's theory. Furthermore, he accounts for the temporal asymmetry of scientific explanation of events by pointing out that causation is asymmetric. We do not explain events by reference to events which succeed them because all scientific explanations are causal explanations and a cause always precedes its effects.

The outline of the Hempel and Salmon theories of explanation shows that Salmon's theory has certain advantages over Hempel's theory. In the next sections I shall examine the consequences of these theories for explanation in physical science and we shall see whether the account of physical explanation derived from Salmon's theory is preferable to that derived from Hempel's theory.

2. PHYSICAL SCIENCE

The overall aim of physical science is the formulation and application of laws which describe relations among quantitative properties of physical systems. These relations may be called quantitative patterns. A quantitative property is one which admits of degrees, or of variable magnitude, and its possession by physical systems enables these systems to be ordered by a suitable class of relations (Ellis [1966], pp. 25–32). Quantitative properties are represented by quantitative concepts or quantities. A quantity, such as mass, is a function which assigns real numbers to physical systems as measures of the magnitudes of the quantitative properties exhibited by the system. For example the quantity *mass* may be expressed

$$M: \Sigma \rightarrow R^+ \tag{4}$$

where Σ is the set of physical systems and R^+ the set of real numbers.

Physical laws are normally expressed by equations, variables are normally used to express the quantities denoted by physical laws, and the values of these variables designate measures assigned to magnitudes of quantitative properties represented by quantities. There are different sorts of physical laws and so different sorts of equations are used to express these laws. Two examples serve to illustrate this. The first is the ideal gas law

$$PV = nRT \tag{5}$$

where P is the pressure, V the volume, T the temperature and n the number of moles of a gas and R is the gas constant. This law does not describe the way in which the values of the variables P, V, T and n change with time; it states the relations among the values of these variables at equilibrium. The ideal gas law is, therefore, typical of phenomenological thermodynamics of which it is a member. The second example is Schrödinger's equation

$$H\Psi_n = E_n \Psi_n \tag{6}$$

where H is the energy operator, Ψ_n is a set of state functions and E_n is a set of energy values. The Schrödinger equation is the expression in the Schrödinger formalism of a basic quantum mechanical law. In this formalism, functions such as Ψ_n represent the states of physical systems and operators such as H represent quantitative properties. The values of the variables which constitute the states of a physical system which are described by Ψ_n, such as the energy values E_n, are determined by solving an equation derived from (6) by substituting an operator of suitable form for H. The form of the operator must be specified if (6) is to yield results. In this respect the Schrödinger equation is similar to Newton's second law of motion which requires specification of the force function before it can be applied. The energy operator H is a differential operator, and hence (6) gives rise to differential equations. In physical science relations among dynamical properties are usually expressed by differential equations.

The application of physical laws to particular physical phenomena requires calculations. For example, if the values of the pressure and volume of one mole of gas are known, then the temperature of the gas is given by

$$T = PV/R \qquad (7)$$

This calculation is quite straightforward. The value of T can be obtained exactly simply by multiplying the values of P and V and dividing by R. However, the application of laws such as the Schrödinger equation is more difficult. In order to apply (6) it is necessary to find the functions of Ψ_n which satisfy the equation. It happens that exact (i.e., analytic) solutions can only be found if the physical system under investigation comprises less than three interacting bodies. If the physical system is composed of three or more interacting bodies, then functions which exactly solve the particular Schrödinger equation which describes the system are not available. The form of the Schrödinger equation which describes three interacting bodies is more complex than that of the Schrödinger equation which describes two interacting bodies. There is no reason in principle why solutions to the more complex equation cannot be found, it just happens that these solutions are not numbered among the stock of functions known to mathematicians.

If the scope of quantum theory were to be restricted to those systems for which exact solutions for Schrödinger's equations are available, then the theory would have very few applications. Fortunately, worthwhile results can be obtained for complex systems by using *approximate* methods. These methods enable us to determine solutions which are approximations to the unknown exact solutions. For example, a problem which involves three interacting bodies may be decomposed into a set of three two-body problems; each of these has an exact solution. The solutions to the two-body problems are then used as models for constructing approximate solutions to the three-body problems. This is a stepwise process which involves the introduction of finer and finer perturbations of the two-body problems. The technique which employs exactly soluble models in this way is known as perturbation theory.[1] The need for approximate methods is not confined to quantum theory. In any branch of physical science in which differential equations express laws, approximate methods are required for the applications of these laws to all but the simplest physical systems.

The widespread use of approximations is a significant feature of the application of laws in physical science. There is another important feature of these applications which should be discussed; this can be illustrated by the example in which the value of the temperature of a gas is calculated from the ideal gas law and the antecedent values of pressure and volume. In practice, the values of P and V are obtained from measurements. However, measurement does not yield individual real number values as measures of the magnitudes of quantitative properties; rather, measurement gives *intervals* of numbers. There are several reasons why we cannot obtain single real number values from measurements.

First, it is impossible to individuate the members of the continuum on a scale, hence it is not possible to construct an instrument on which real numbers are distinguished. On any given measuring instrument there is a certain minimum difference detectable. This is known as the *resolution* of the instrument, and it sets an upper limit on the accuracy of results which can be obtained using the instrument. For example, suppose we can detect differences of one hundredth of a degree Kelvin using a certain thermometer, then the most accurate readings which can be obtained using this instrument have the form

$$T = x \pm 0.005°K \tag{8}$$

Differences of less than $0.005°K$ cannot be detected.

Second, there are physical reasons why measurements cannot be made with unlimited accuracy. The interaction between the measuring instrument and a physical system of interest is governed by the laws of quantum theory. Hence there are limitations on the amount of information which can be transferred to the instrument and hence on the precision of the readings which can be taken from the instrument. Thus, the values of variables which are determined by measurement are intervals of numbers, not individual numbers, and the size of the interval is determined, as in (8), by the error in the reading.

Although single real number values cannot be obtained by measurement, it is assumed that physical systems actually exhibit magnitudes of quantities which have single real number values at given times.[2] The interval of values indicates the limitations of what we can know about physical systems, it does not indicate that they exhibit magnitudes which are precisely quantified by intervals of numbers. However, it may be possible to write down an interval of numbers which contains the actual (single real number) value of a variable exhibited by a physical system on the basis of measurement. If this happens, then the statement in question will be said to be true. Although we can never know that a statement of this kind is true, we can have grounds for believing it to be true if the results obtained by the measurement techniques have been reliable in the past.

Errors are propagated and compounded by calculation, and so values of variables which are determined by calculations from laws and antecedent conditions are also intervals of numbers. For the generalised product

$$u = cx^l y^m z^n \ldots \tag{9}$$

the maximum percentage error is given by the rule

$$Eu/|u| = |lE_x/x| + |mE_y/y| + |nE_z/z| \tag{10}$$

where E_x is the maximum percentage error in x, etc. The modulus indicates that successive terms must be added (Lyon [1970], pp. 38–40); this is because we must anticipate the worst possible case when dealing with errors. The rule for error propagation expressed by (10) shows how errors are compounded by calculations. For example (7) and (10) give

$$E_T/T = E_P/P + E_V/V. \tag{11}$$

The percentage error in the calculated value of the temperature of the gas is greater than the percentage errors in the observed values of the pressure and volume. If it is possible to measure the values of P, V and T with roughly the same accuracy, then the calculated values of these variables will be less accurate than the measured values. This means that the intervals of numbers which represent the values of these variables determined by calculation will be larger than the intervals determined by measurement.

In this section some aspects of physical science have been mentioned. In particular, the widespread use of approximative methods in solving the equations which express physical laws and the fact that the values of variables obtained from measurement are intervals of real numbers, and not individual real numbers, which are compounded by calculation are noteworthy features of physical science. No attempt has been made to interpret these applications as physical explanations; in order to do this it is necessary to adopt some theory of explanation.

3. PHYSICAL EXPLANATION

We have seen that physical science is mathematical science, that is, that physical science is concerned with properties of physical systems which can be quantified. This suggests that the questions which are answered by physical science have something like the following form: "Why does physical system P exhibit property Φ to degree r?" It is assumed that physical systems possess properties such as Φ, and the questions which are of interest in physical science are why these systems exhibit these properties to the degree to which they do in fact exhibit them. We must now ask under what conditions these questions are answered.

Hempel believes that explanations in physical science conform closely to the D-N model. For example, he writes:

Some scientific explanations conform to the pattern [the D-N model] quite closely. This is so, particularly, when certain quantitative features of a phenomenon are explained by mathematical derivations from covering general laws . . . ([1966], p. 52).

On the Hempel theory, the derivation (calculation) of the value of the temperature of a gas from antecedent conditions which record the value of the pressure and volume of the gas and the ideal gas law is a scientific explanation of the value of the temperature as it is described by the explanandum. On the assumption that the ideal gas law and the statements of antecedent conditions are true,[3] then the explanans clearly satisfies the first three conditions of the D-N model. The calculation of the value of the temperature is carried out in accordance with deductive rules, and so it appears that the explanandum, which is a statement which records the value of the temperature determined by observation and measurement, is deducible from the explanans provided that it is true. Furthermore, given that the explanandum is deducible from the explanans, then the explanans clearly leads us to expect the temperature to have the value which it does in fact have. The solution of a differential equation and the determination of the calculated value of a variable from this solution also takes place in accordance with deductive rules, and so there seems to be no reason why applications of laws which are more complicated than the straightforward determination of the value of the temperature of a gas should not be interpreted as explanations on the Hempel theory. However, when we examine the consequences of the two features which mark applications of physical laws which were mentioned in the previous section, we see that the Hempel account of physical explanation is not really satisfactory.

It is useful here to introduce some terminology. Let E be an explanandum which records the value of a variable obtained from measurement and let (c, d) be the interval of numbers recorded by E. Let P and A be a set of laws and a set of antecedent conditions respectively; let D be the statement deducible from P and A which describes the same value of the same variable as E and let (a, b) be the interval recorded by D. The derivation of D is an application of P to the particular circumstances described by A. In most of the examples discussed by philosophers of science who accept some deductivist model of scientific explanation it is not necessary to distinguish the explanandum and the statement derivable from the explanans, because these two statements are identical. But it is most unlikely that D and E will be identical. For this to be the case, it is necessary that (a, b) is identical to (c, d): if (a, b) is not identical to (c, d) then D is not the same statement as E. The interval (a, b) is determined from the intervals recorded in A in accordance with the relations specified by P. It is most unlikely that an interval determined in this way would be precisely the same size as (c, d), which is determined in a quite different way.

If E is not identical to D, it does not follow that E is not deducible from P and A. If D entails E, then E is deducible from P and A; the conditions under which D entails E do not require that D and E are identical. Thus, from the point of view of the Hempel theory, it is important to find out the conditions under which D entails E. We have seen that the truth of statements such as D and E depends upon whether actual (single real number) values lie within the intervals recorded, hence the entailment relations between D and E depend upon the relation between these intervals. To see this, consider the following three possibilities:

(i)   All the members of (a, b) are members of (c, d).
(ii)   Some members of (a, b) are not members of (c, d).
(iii) No members of (a, b) are members of (c, d).

In case (i), since all the members of (a, b) are members of (c, d), the actual value of the variable must be a member of (c, d) if it is a member of (a, b). In this case D entails E. In case (ii), since some members of (a, b) are not members of (c, d), then if the actual value is a member of (a, b) it may or it may not be a member of (c, d). In this case D and E are consistent, but D does not entail E. In case (iii), (a, b) and (c, d) have no common members, and so if the actual value lies within (a, b) it cannot lie within (c, d). In case (iii), D entails the negation of E. Summarising the entailment relations, we have

(i) D entails E.
(ii) D and E are consistent.
(iii) D entails the negation of E.

On the Hempel theory, it is only application of laws which conform to (i) which are interpreted as physical explanations, because it is only in this case that the fourth condition of the D-N model is satisfied.

Applications of physical laws which conform to (i) are much less common than those which conform to (ii) or (iii). This is due to the fact that errors of measurement are compounded in calculation and to the widespread use of approximations. We have seen that the error in the calculated value of a variable is greater than the errors in the antecedent values. Thus, if the errors recorded by E and A, which are the results of measurements, are of the same order of magnitude, then the error recorded by D will be greater than that recorded by E. This implies that (a, b) is larger than (c, d) and hence that D does not entail E. It is possible that the accuracies of the measurements recorded in A are very much greater than those recorded in E, and so, even allowing for the amplification of error by the calculation, it is possible that (a, b) is smaller than (c, d) and hence that an example of (i) may be obtained.[4] There is, however, no reason to suppose that this combination of circumstances is at all common. On average, it is reasonable to suppose that the accuracies of the readings recorded in A and E will be about the same, and hence that examples of (i) are exceptional.

In addition to error combination, the use of approximations tends to increase discrepancies between (a, b) and (c, d). Whereas the combination of errors tends to increase the size of (a, b), the use of approximations tends to shift (a, b) away from (c, d). If D and E are both true, then (a, b) and (c, d) must overlap because they have at least one common member, namely, the actual value of the variable under consideration. If A, P and E are true and if the derivation of D is sound, then (a, b) and (c, d) must overlap. An application of P which satisfies these conditions will therefore conform to (i) or (ii). However, if an approximation is used, then D is not deduced from P and A. The derivation of D is complex; it involves considerations determined by the approximative technique employed. Since D is not derived from only true premises, there is no guarantee that D is true even if P and A are true. Hence, when an approximative technique is employed for the derivation of D, there is no guarantee that (a, b) and (c, d) will overlap. It is in this sense that the use of approximations tends to shift (a, b) away from (c, d). We have noted that applications of laws which are expressed by differential equations require the use of approximations in most instances.

There are, therefore, good reasons for concluding that Hempel's theory of scientific explanation considerably restricts the possibilities for the explanation of particulars by deterministic laws in physical science, because it requires that applications of laws must conform to (i) to be interpreted as physical explanations. However, examples of (i) are uncommon. Although it does not follow that Hempel's theory is false because it restricts the scope of physical explanation, we should prefer a more liberal account. In Section 1, Salmon's theory of explanation was seen to have certain advantages over the Hempel theory. Let us see whether it gives a more satisfactory account of physical explanation.

Salmon's theory does not include the fourth condition of the D-N model, namely, that the explanandum must be deducible from the explanans, hence on this theory applications of physical laws which conform to (ii) and (iii) are not ruled out as explanations because they do not satisfy this condition. At first glance, Salmon's theory is more promising than that of Hempel.

According to Salmon ([1978c], p. 699), if we wish to explain a particular event we begin by assembling factors statistically relevant to its occurrence. For example, suppose we wish to explain an increase in pressure of a gas in a closed vessel (i.e., a fixed volume of gas). The gas law specifies that the value of the pressure of the gas is proportional to the values of the temperature, volume and the number of moles n

$$P = nRT/V \tag{12}$$

Thus, the gas law tells us that changes in n, T and V are statistically relevant to changes in P. On the S-R model (Salmon [1970b], p. 79), a deterministic law is interpreted as the limiting case in which the probability of one event given that another event has occurred is unity. The gas law can, therefore, be interpreted as specifying statistical relevance relations. Let us suppose that, prior to our observation of the increase in pressure of the gas, we had observed that a certain amount of heat had been imparted to the vessel and that this has raised the temperature of the gas. The gas law tells us that a change in temperature is statistically relevant to change in pressure, and so the change in temperature is a relevant factor. Furthermore, if n remains constant, then this change is the only relevant factor for the increase in pressure in this instance.

The assembly of the statistically relevant factors is only the first stage in the explanation of the change in pressure of the gas. Unless the statistical relevance relations in question can be given a causal interpretation, there is no scientific explanation. It is well known that the gas law can be reduced to the kinetic theory. This reduction involves giving mechanical interpretations to the thermodynamic state variables P and T. For example

$$T = mv^2/Nc \qquad\qquad\qquad\qquad\qquad (13a)$$
$$P = Nmv^2/3V \qquad\qquad\qquad\qquad\qquad (13b)$$

Where N is the number of molecules of gas, c is a constant, m is the mass and v the average velocity of a molecule of gas and V is the volume of the vessel. On this interpretation, both T and P are proportional to the average kinetic energy of the molecular constituents of the gas. An increase in the temperature of the gas is equivalent to an increase in the average kinetic energy of the molecular constitutents and hence to an increase in the pressure of the gas. The changes in P and T are, therefore, seen to be due to the heat gained by the gas system. The number of moles n of gas and the gas constant R can be expressed in terms of N and c, and so the gas law can be deduced from a set of premises which include (13). The kinetic theory of gases is a causal theory in Salmon's sense ([1978c], p. 698). Thus the reduction of the ideal gas law to the kinetic theory is the second step necessary for the explanation of the change in pressure of the sample of gas.

In the example which has just been discussed, it is certainly plausible to claim that the explanation of the change in pressure requires a causal explanation and that the inference of a statement of the value of the pressure from the gas law and suitable antecedent conditions, which counts as an explanation on Hempel's theory, is not sufficient. However, suppose we wish to explain why a sample of gas exhibits a particular *set* of values v, w, x, y for P, V, n and T. That is, we are not concerned to explain *changes* in the values of these variables, but why they exhibit certain *equilibrium* values. A possible explanation is that these values are in accordance with the gas law, namely that

$$vw = xyR \qquad\qquad\qquad\qquad\qquad (14)$$

This, v, w, x, y is a combination of values which is allowed by the law, whereas, for instance, v, w, x, z, is not. The example does not fit into the Salmon theory; because there is no *event* to be explained, there is no need to introduce causes. The ideal gas law can, of course, still be given a causal interpretation, but this does not add anything to the explanation of the equilibrium values of the variables in question. In general, there seems no reason why we should not ask for explanations in physical science of equilibrium values of variables.

Phenomenological thermodynamics is the theory of physical science which describes equilibria. This theory is believed to be replaceable by statistical mechanics which is a causal theory in Salmon's sense, and hence the equilibria described by phenomenological thermodynamics are interpreted as dynamic equilibria. This suggests that a causal explanation may be possible for sets of equilibrium values, and so the objection to Salmon's theory mentioned in the previous paragraph loses much of its force. There is, however, a more serious objection to the Salmon theory. This is that it rules out applications of the laws of quantum theory as explanations because quantum theory is not a causal theory. If it is true that Salmon's analysis of cause implies that there cannot be explanations in quantum theory, then this is a damaging consequence. We may accept that phenomenological thermodynamics does not give explanations because this theory is replaceable by one which does — statistical mechanics — but quantum theory is the most advanced and successful physical theory and we would be most unwilling to accept that it does not provide us with explanations.

In Section 1, Salmon's analysis of cause was briefly discussed. On this analysis, a causal process is distinguished from a pseudo-process by its ability to transmit a mark from one spatio-temporal region to another. The mechanism by which a mark is transmitted from A to B is simply that it is at the intervening points between A and B at suitable temporal instants:

> The transmission of a mark from point A in a causal process to point B in the same process *is* the fact that it appears at each point between A and B without further interactions ... The basic thesis about mark transmission can be stated as follows: *A mark that has been introduced into a process by means of a single intervention at point A is transmitted to point B if it occurs at B and at all stages of the process between A and B without additional interventions* (Salmon [1977f], p. 221).

> The at-at theory of mark transmission provides, I believe, an acceptable basis for the mark method, which can in turn serve as the means to distinguish causal processes from pseudo-processes (Salmon [1977f], p. 223).

The second quotation shows that the at-at theory of mark transmission is basic to Salmon's theory. However, it appears that the processes described by quantum theory are not causal processes according to the at-at criterion. For example, an electron which passes from point A to point B does so without having a definite trajectory. If it does not have a definite trajectory, then it cannot appear *at* each point between A and B. There are a number of well-known illustrations of this. For instance, the energy of an electron in a bound state in an atom is quantized; it can only take on integral values of energy. The energy is proportional to the separation between the electron and the nucleus,[5] hence the electron can only occupy certain regions in the neighbourhood of the nucleus. These are the so-called orbits or stationary

states of the electrons. There is zero probability of the electron being in some
other region, since this would correspond to a non-integral value of the
energy. This leads to a very puzzling state of affairs. Electrons can emit and
absorb radiation – but only in integral amounts – and so they jump from
one stationary state to another. But there is zero probability of their occupying points in between the allowed orbits. As long as we retain the picture of
the electron as a classical particle this process is incomprehensible because
classical particles do not jump from one place to another without occupying
some intervening regions. Even if it is possible in some way to picture
electronic transitions – and this is not necessary for their explanation [6] –
this cannot involve the electron following a definite trajectory from one
orbit to another because this violates the laws of quantum theory. The
process constituted by an electron transition within an atom is not, therefore,
a causal process on the at-at criterion because the electron is not at each
point on a trajectory between the two orbits in question.

The reason why Salmon's analysis does not allow that the processes which
are typically described by quantum theory are causal is that the at-at account
of mark transmission is essentially a classical account. Classical particle
mechanics defines the state of a particle as comprised of the values of position and momentum of the particle. Position and momentum are represented
by (continuous) functions in classical mechanics; hence a definite trajectory
is computed for a particle as it moves from A to B under forces from the
equation of motion of classical mechanics. Thus, the function which specifies
the path of the particle determines that the particle is at a particular point
at a particular time. This is completely in accord with the at-at theory.[7]

In quantum theory, quantitative properties are represented by operators
as in (6) and not by continuous functions. Although Ψ is a function, it does
not directly describe quantitative properties. On the Born interpretation

$$\Psi\Psi^* \, dr = p(r) \tag{15}$$

where Ψ^* is the complex conjugate of Ψ, dr is a volume element, and $p(r)$
is the probability that a system in a state described by Ψ is located within
dr. In general, applications of the laws of quantum theory give probabilities
that the system under discussion exhibits certain values of variables. Quantum
mechanical systems can, therefore, behave in a manner which is inherently
discontinuous.

The Born interpretation shows that even when an electron is not in a
bound state, its motion can be discontinuous. If an electron is not in a bound
state, then $p(r)$ is not necessarily non-zero for any volume element dr. Thus,

if an electron is emitted from a source at A and is absorbed by a target at B, then quantum theory does not tell us that the electron follows any definite trajectory from A to B. It is possible that the electron is 'at' one point at one instant in time and 'at' a non-neighbouring point at the next instant. In which case, the transfer of the electron from A to B is not a causal process on the Salmon theory.

There are two ways in which Salmon's theory might be modified to escape the charge – implied by the above comments – that it is unduly restrictive because it rules out the possibility of explanations in quantum theory. First, it could simply be denied that quantum theory provides us with explanations. In which case the fact that the Salmon view apparently rules out explanations in quantum theory becomes a point in its favour. However, I do not recommend this alternative, nor do I think it is acceptable to Salmon. Quantum theory enables us to account for many surprising phenomena, including the distribution of blackbody radiation and the line spectra of atoms; furthermore, this theory applies to every physical process involving matter and radiation. To deny that quantum theory provides explanations is to deny that the most successful and most widely applicable scientific theory provides explanations. This conclusion would be welcomed by an instrumentalist who claims that scientific theories are mere predictive devices, but it would not be welcomed by a realist, such as Salmon, who holds that the aim of science is to explain. Consequently the first alternative should be discarded.

We have seen that the reason why Salmon's account of explanation rules out explanations in quantum theory is that the at-at theory is a classical theory of motion. Thus, if we are to have explanations in quantum theory, the at-at theory must be rejected and a new basis for distinguishing causal processes from pseudo-processes must be found. However, it would be a mistake to replace the classical at-at theory with some quantum mechanical account of mark transmission, because this would define scientific explanations as explanations which are consistent with a particular scientific theory. This is precisely what is wrong with the Salmon view as it stands: it identifies scientific explanations with the explanations of classical physics. It is possible that quantum theory will be replaced by a deterministic theory. An account of scientific explanation which defines mark transmission, and hence causal process, by reference to quantum theory will not allow explanations in terms of the new deterministic theory. If the distinction between causal and pseudo-processes cannot be drawn except by reference to a particular theory, then the concept of a scientific explanation will be made relative to that particular theory. This consequence is unacceptable because the class of scientific explanations is wider than the class of explanations given by a particular theory.

As a possible solution to the difficulty which I have identified in the Salmon theory of explanation, I suggest the following: a causal process is one which is governed by scientific laws (theories), whereas a pseudo-process is not. The advantage of this suggestion is that it does not refer to any *particular* laws or theories.

4. CONCLUSION

The main purpose of this paper has been to examine the consequences of the theories of scientific explanation of Hempel and Salmon when they are used to provide accounts of the explanation of individual events in physical science. In Section 2 some features of the applications of physical laws were mentioned, and in Section 3 the accounts of physical explanation of the two theories were discussed by examining the way in which the theories interpreted applications of these laws as explanations. It was suggested that both accounts of physical explanation are restrictive.

On Hempel's theory of explanation, scientific explanations are arguments. In the case of the explanation of individual events by deterministic laws, these arguments have a logical structure which is specified by the D-N model. However, we have seen that applications of physical laws which exhibit this structure are most uncommon, and hence the Hempel theory restricts the possibilities for physical explanation. Although Salmon's theory does not entail that the explanandum is deducible from the explanans, it also seems to impose limitations. Unless a law is, or is reducible to a causal law, an application of the law cannot give rise to scientific explanations. This limitation rules out certain examples which we would not wish to count as explanations, but it also appears to rule out explanations in terms of quantum mechanical laws, which are not causal in Salmon's sense.

Some attempt was made to determine why the Salmon theory is restrictive, and it was suggested that the at-at theory of motion is responsible. Finally, as an attempt to resolve the difficulty of Salmon's theory, it was suggested that causal processes are processes governed by laws.

NOTES

† Added in proof: This paper was written in 1979. Since that time van Fraassen's book ([1980]) has appeared in which it is mentioned that Salmon's causal account does not cover certain quantum mechanical processes.

1 Some account of perturbation theory and other approximation techniques is given in most books on quantum chemistry. For example see Atkins [1970], Chap. 5.

2 It is not true that all systems can exhibit a continuous spectrum of values. In some circumstances the values of variables are quantized. However, in such circumstances it is still not possible to determine these values precisely because of physical effects such as Brownian motion.

3 The ideal gas law is false. However, I treat it as if it were true in the discussion which follows.

4 The accuracy of measurements changes as measurement technology progresses. This means that smaller intervals of number are recorded by statements of measurements as measurement technology progresses. I assume that we never prefer a less accurate measurement technique to an equally reliable more accurate technique. Thus, I assume that we place limitations on the size of (a, b) and (c, d) and that arbitrarily large (c, d)s in physical explanations are not acceptable.

5 The form of the energy operator for an electron in a hydrogen atom is

$$H = -(\varrho^2/2\mu)\nabla^2 - e^2/4\Pi \, e^0 r$$

where r is the separation of the electron and the nucleus.

6 Most physical scientists do not believe that the phenomena described by quantum theory can be pictured in classical terms. For example, according to Dirac: "In the case of atomic phenomena no picture can be expected to exist in the usual sense of the word picture, by which is meant a model functioning essentially on classical lines" ([1958], p. 10).

7 Salmon has, in fact, used the concept of a function to explicate the at-at theory as it is applied to the solution of Zeno's paradox of the flying arrow ([1975a], p. 41). The at-at theory can still be used to explain the motion of the arrow if space is taken to be discontinuous, but it does not account for discontinuous motion in continuous space.

FURTHER REFLECTIONS

Clark Glymour pays me a kind — and probably undeserved — compliment when he says that I am really a son of the Enlightenment. There is, however, one sense in which it is certainly true. Early in my studies of philosophy I encountered Hume's *Enquiry Concerning Human Understanding*, and it is fair to say that I never recovered from the experience.

Hume must be considered, in retrospect, one of the towering figures of the Enlightenment, even though he was not fully recognized as such at the time. This subtle and penetrating intellect perceived basic philosophical problems associated with two fundamental concepts — *induction* and *causality* — and presented them with unparalleled clarity. The combination of critical acumen and stylistic elegance proved irresistible to me. The significance of these problems was not, I think, fully appreciated until the present century, and even now, there are many philosophers who prefer to take evasive action rather than meeting them head-on. I disagree emphatically. It seems to me that we cannot provide adequate accounts of scientific inference or scientific explanation without coming directly to terms with Hume's classic critiques of these two concepts. The sins of the father are visited upon succeeding generations.

The essays in this volume take up a number of issues which have concerned me during the past thirty years. Many views are expressed with which I agree, and many with which I disagree. In these comments, I shall not, however, attempt to detail all of the points either of agreement or of disagreement. Instead, I shall try to address what I see as the "larger issues". I hope, in this way, to do my bit to contribute to our common enterprise — deeper understanding of the fundamental philosophical problems with which we have all struggled.

INDUCTION AND PROBABILITY

A rather large portion of my work has been devoted to problems concerning induction and probability. My doctoral dissertation was in that area, and for several years my meagre list of publications contained no item in any other area. I started out with the conviction that the problem of justification

Robert McLaughlin (ed.), What? Where? When? Why?, 231–280.
Copyright © 1982 by D. Reidel Publsihing Company.

of induction is a meaningful and fundamental one, and that Reichenbach had furnished an adequate solution. A great deal of my effort was devoted to attempts to refute the thesis — which enjoyed enormous popularity in the 1950's, and continues to find considerable favor — that this so-called problem is not a real problem which demands solution, but rather, a pseudo-problem which requires dissolution (Salmon [1957a]). I still think that this approach to the problem of justification of induction is unsound, and I have tried on many occasions to say why. The failure of dissolutionists to take account of any of these arguments has been a source of distress (Salmon [1978a]).

At first I thought that Reichenbach's inductive theory stood in need of completion rather than correction. It seemed that Peirce's problem of the short run had to be faced, but that it could be resolved by means of a pragmatic justification similar in basic respects to Reichenbach's pragmatic justification of induction (Salmon [1955]). It also appeared that Reichenbach's rather cryptic remarks about the use of Bayes's theorem in handling the probabilities of hypotheses required considerable clarification (Salmon [1967a], chap. VII). It began to dawn upon me, however, that Reichenbach's justification of induction itself suffered from a basic flaw. His appeal to descriptive simplicity to select his own rule of induction from the infinite class of asymptotic rules could not be valid. In my [1957b] I tried to show just how critical the problem was for his pragmatic justification of induction.

For several years, I attempted to repair the difficulty in Reichenbach's approach by finding grounds for rejecting as irrational large subclasses of the class of asymptotic rules. The hope was that all but one of the asymptotic rules could thus be dispatched, leaving only the rule of induction by enumeration as an acceptable asymptotic rule. If that could be done, then Reichenbach's justification for the use of asymptotic rules would thereby be transformed into a justification for a unique member of that class. This approach, it seems to me, enjoyed partial success, for the "normalizing conditions" and the "criterion of linguistic invariance" did provide good grounds for eliminating many important types of asymptotic rules. For example, I believe that the normalizing conditions eliminate a large subclass of Reichenbach's asymptotic rules (Salmon [1956b]), and that the criterion of linguistic invariance disqualifies all of the inductive methods in Carnap's continuum (Carnap [1952]) save for the "straight rule" (Salmon [1961a]). For a brief time I thought that these two requirements would do the whole job (Salmon [1963b]), but I was soon disabused of this bit of over-optimism. The crucial development occurred at a 1965 conference when Ian Hacking commented upon my [1968b]. In an appendix to the published version of his comments

(Hacking [1968]), he presented three criteria which are necessary and sufficient to select the rule of simple enumeration from the entire class of asymptotic rules. It looked easy to justify one of these criteria, and it seemed a moderately tractable problem to find a justification for another. I had no clue as to how to deal with the third.

It is often good strategy to walk away from a problem which appears intractable, and to come back after a time, hoping for a fresh idea. Since Hacking's critique was published, my main efforts have been directed toward other issues, but whenever I re-approached the problem of justification of induction, fruitful fresh ideas have not been forthcoming. Many people have concluded, I fear, that Hacking demonstrated the impossiblity of providing a pragmatic vindication of induction. That assessment is incorrect. What Hacking did was to define the problem in clear and precise terms. For this we owe him a large debt of gratitude.

Although much has been written since 1968 on inductive logic, I have not seen anything which adequately handles the problem of justification of induction (see Salmon [1978a]). By and large, philosophers seem greatly relieved at not feeling obliged to deal with this problem. John Clendinnen has not shared this popular view; instead, he has diligently pursued the problem. He has, I believe, provided *the fresh idea* needed to make significant progress toward a solution, if not to achieve a fully satisfactory result. Thus, when I read his essay, "Rational Expectation and Simplicity", I wrote to tell him that — because of his efforts — I now feel more optimistic than I had for many years about the prospects for solving this deep and difficult problem. The fact that his paper is the lead essay in this volume is a source of great joy to me. I now believe it is possible, through the use of Clendinnen's basic insight, to justify each of the three criteria enunciated by Hacking. However, in these relatively brief and informal "reflections" I shall confine myself to some general remarks, reserving the technical details for another occasion.

The status of the normalizing conditions seems secure, for violations can be straightforwardly construed as logical contradictions. In presenting the criterion of linguistic invariance, I have always pointed up the opportunities for contradictions to arise in case of its violation, since I hoped that this criterion could be furnished with as strong a rationale as the normalizing conditions have. At the same time, I have emphasized the fact that the troubles which occur when it is violated arise out of the fact that the inductive evidential relation is made a function of some *arbitrary* features of the choice of a descriptive language. Suppose, for example, that I draw three balls from an urn, and find that two of them are red. Living in a bilingual community, I might expect one observer to describe these results in Spanish while another observer describes them in English. Both languages are descriptively adequate to formulate this evidence, and there is no doubt that the statements are semantically equivalent. If both observers use the same inductive rule R, and if they get different posits for the value of the probability of red on draws from the urn, it is evident that an inconsistency is present in the two posits taken together. The mathematical calculus of probabilities stipulates that values of probabilities are unique. Clearly, however, no one is forced into a contradiction if everyone sticks to only one language. Nevertheless, if this situation is allowed to exist, we must protest that the choice between English and Spanish is, in this context, inductively arbitrary. This, I take it, is the tack Clendinnen would advise us to follow. I believe he is right.

There is a further consideration which suggests that it is arbitrariness, rather than risk of contradiction, which is troublesome in such circumstances. In his earlier works (esp. [1950] and [1952]), Carnap used features of the language in which evidence and hypotheses are formulated as a basis on which a priori probabilities are assigned. This made his inductive logic vulnerable to the criterion of linguistic invariance.

In some of his later works, he reformulated his system of inductive logic in such a way that the a priori measures were assigned in terms of properties rather than predicates (see Salmon [1967b]). Although Carnap was not motivated by consideration of the criterion of linguistic invariance, it is evident that such a move would undercut the criterion. Yet, it is equally evident that the incorporation of the properties, rather than the predicates which we use to name them, into our basic inductive rules is just as arbitrary as is the direct relationship of the rule to the language in which the properties are described. Suppose we are dealing with balls which may be either red or green, while no other colors are present in the urn. Suppose that a particular observed frequency of red yields a particular posit for the value of the probability of red in the whole population. If the same observed frequency of green in a sample of the same size were to yield a lower posit for the value of the probability of green in the whole population, that would be just as arbitrary as it would be to say that predicates beginning with "r" in English are to be given higher inductive weight than predicates beginning with "g" in English. Considerations of this sort led me to argue for a "criterion of statistical invariance" which runs quite parallel to the criterion of linguistic invariance (Salmon [1967b] 733–735). As far as I can see, there is no prospect of showing that violations of this criterion run any risk of resulting in logical contradiction. If this criterion is defensible, it is on the ground that violations lead to intolerable arbitrariness in our inductive inferences. Clendinnen's approach thus provides grounds for both of these invariance criteria.

The same basic idea can be applied to Reichenbach's attempt to justify his rule of induction. In *Experience and Prediction* [1938], he argued that, while all of the asymptotic rules will lead to correct posits of limiting frequencies if they exist, his rule of induction is preferable because it will achieve correct results sooner than the others (p.355). In *The Theory of Probability* [1949], he recognized that no such conclusion can be established, and he looked for a different basis for preferring his rule of induction to all of the other asymptotic rules. He claimed that this selection could be justified on grounds of descriptive simplicity (pp. 475–476).

It was Reichenbach himself who had clearly enunciated the distinction between "inductive simplicity" and "descriptive simplicity" ([1938], Section 42). In certain contexts, we prefer a simpler hypothesis to a more complex hypothesis because we have some reason – experience in that particular scientific field perhaps – to believe that the simpler hypothesis is more likely to be true. In such cases, the hypotheses are not semantically equivalent to one another, though each would presumably be consistent with all available relevant data. This is the kind of situation in which inductive simplicity is invoked.

In other contexts, we have two or more hypotheses which are semantically equivalent to one another. An important example involves Reichenbach's thesis – discussed at some length in the essays by Nerlich and Saunders/Norton – concerning alternative admissible definitions of simultaneity in special relativity. These different definitions, based upon different values of ϵ, lead to different descriptions which are factually equivalent to one another. However, we choose one of these definitions – that based upon $\epsilon = \frac{1}{2}$ – because it leads to descriptions which are enormously more simple and straightforward than those based upon other definitions. No factual error would be introduced by adopting a different definition, Reichenbach claimed, but we would make our lives needlessly complex by adopting one of the non-standard alternatives. Since two equivalent descriptions are either both true or both false, there is no possibility of selecting one over another on grounds of its greater likelihood of being true. In such situations we invoke descriptive simplicity for practical or aesthetic reasons.

There are two peculiar things about Reichenbach's appeal to descriptive simplicity as a ground for selecting the simplest inductive rule. First, Reichenbach was one of the first philosophers to make a clear distinction between the problem of justifying a *rule* and the problem of providing support for a *statement* such as a principle of uniformity of nature. His approach to the problem of induction in terms of a pragmatic justification depends crucially upon recognition that a rule is being justified. Nevertheless, both inductive simplicity and descriptive simplicity apply to a choice among statements, not a choice among rules. Neither of these two types of simplicity is directly applicable to the problem at hand.

Second, even if we overlook this problem, and consider the basis Reichenbach offered for his appeal to descriptive simplicity, it appears to be totally unconvincing. Reichenbach says that, since all of the asymptotic rules converge, they all give the same results in the long run. Even if they *are not* statements, they *lead to* equivalent factual statements in the long run. This argument is altogether unsatisfactory, however, for as I showed in [1957b], the class of asymptotic rules is, for all human applications, as divergent as it is possible for a set of rules to be. Pick *any* sample size n (where n is a positive integer), *any* observed frequency m/n (where m is any integer such that $0 \leqslant m \leqslant n$), and *any* probability value p (where p is any real number such that $0 \leqslant p \leqslant 1$). Within the class of asymptotic rules, there is at least one which will license us to posit the probability p upon the evidence of an observed frequency m/n in a sample of size n. The class of asymptotic rules therefore holds possibilities of unlimited arbitrariness in our inductive inferences.

Once we have recognized that the problem which confronts us is one of selecting among rules rather than among statements, it should be obvious that we need a type of simplicity different from the two types Reichenbach delineated. Let us call it *methodological simplicity*. To invoke methodological simplicity is to choose a simpler rule in preference to more complex alternatives which happen to be available. To say that we can select a rule from a large class of rules on the basis of methodological simplicity does not, of course, provide any justification for that practice. That is the fundamental problem we must face. The basic idea is to show that the adoption of a more complex rule would involve arbitrary considerations, and that such involvements should be avoided. The appeal to methodological simplicity, with its concomitant avoidance of arbitrary features of our rules of inference, seems intuitively appealing, but we must say more. We must try to find arguments which go beyond mere appeal to intuition to support our intuitive judgment.

Let me attempt to illustrate the sort of approach which I find promising. Asymptotic rules, it will be recalled, have the following form: if m/n is the observed frequency with which an attribute has occurred within a sample, then we posit that the probability (long run frequency) of that attribute in the pertinent reference class is $m/n + c_n$, where c_n is some function which converges to zero as n goes to infinity. Now suppose that we are dealing with a mutually exclusive and exhaustive set of attributes B_i, and suppose that c_n is a function of i, where i can be taken as an index either of the predicate "B_i" or of the attribute B_i. In either case, the appearance of i as an argument of c_n would introduce an intolerably arbitrary element into our rule of inference. We can argue as follows. If we have a mutually exclusive and exhaustive set of attributes B_i, then as a matter of simple arithmetic, the sum of the observed frequencies for any given sample must equal one. Likewise, by a basic principle of the probability calculus, the sum of the probabilities must equal one. If, in positing a value for the probability of a particular attribute B_k, we add a positive amount c_n to the observed frequency m_k/n, then we will have to take a like amount away from the observed frequencies of one or more other B's in arriving at posits of their probabilities. On what basis, we must ask, are we to choose which observed frequency to augment and which to diminish?

We might, of course, have additional factual information about the population in question – which might make it possible to assign prior probabilities to various frequency distributions – and if such information is available, it must be taken into account. But this kind of information is not to be incorporated into our basic rules of inductive inference. The kind of rule we are now considering is applicable only in cases in which such additional factual information is unavailable. In these circumstances, any choice of observed frequencies to be augmented or diminished seems utterly arbitrary.

At this point an appeal to a methodological version of the principle of indifference – along the lines suggested by Clendinnen – seems appropriate. Consider two attributes, B_j and B_k. Suppose a rule R_1 is proposed which involves increasing m_j/n by a positive quantity c in order to arrive at a value to be posited for the probability of B_j, and which involves decreasing the observed frequency m_k/n of B_k by the same amount in order to arrive at the value to be posited for the probability of B_k. Consider another rule R_2 which is just like R_1 except for the fact that m_k/n is augmented while m_j/n is diminished. If no argument can be provided to show that R_1 is superior to R_2, or what amounts to the same thing, if every argument for the superiority of R_1 can be paralleled by an equally good argument for the superiority of R_2, then we have no reason to prefer R_1 to R_2. It may turn out that R_1 is, in fact, superior to R_2 – e.g., the values of the posits made according to R_1 may converge more rapidly than those made according to R_2 – but we can have no way, even in principle, of knowing such facts in advance. One might decide between R_1 and R_2 by making a blind guess, by flipping a coin, or by some other arbitrary method, but the choice cannot be justified on any rational basis. In these circumstances, it seems to me, we can reasonably maintain that neither choice would be desirable. If we run through the same sort of argument for each of the attributes B_i with the same result for all, then we can conclude that the posited value of the probability should not depend upon i — that is, i should not be an argument of the function c.

To make c depend upon i would be to make our rule needlessly complex by introducing arbitrary features. It is disadvantageous from a practical standpoint to introduce rules which embody such complexity. The added complexity cannot be shown to have any positive inferential value, and it has the disadvantage of making our rules more complicated to state and more difficult to apply. These considerations suggest that we formulate a new *methodological* principle of indifference:

> For purposes of inductive inference, treat all observed frequencies equally unless there is some positive reason for treating them unequally.

This principle will lead us to eschew rules of inference which introduce "corrective terms" c which are arbitrarily chosen functions of variables which are incapable of being shown relevant to the problem at hand. This principle is not open to the well-known objections raised against either the Laplacian or the Carnapian forms of the principle of indifference, for *it does not yield a priori values of probabilities*. It does not direct us to posit equal values for probabilities on the basis of no observed frequencies at all. It does seem to me to formulate the "valid core" of the principle of indifference which Carnap tried to locate and employ in building his systems of inductive logic. It also seems to capture the type of methodological simplicity to which Reichenbach should have appealed in order to justify his choice of a unique rule of induction.

As I have said above, it is not enough to state a policy of avoiding arbitrariness. The policy also needs to be justified. Something further therefore needs to be said about the irrationality of arbitrary rules. In (Salmon [1957a]) I tried to refute a certain version of the claim that induction can be justified inductively. Instead of arguing that self-supporting arguments are invalid, I tried to show that they are pointless. My strategy was to exhibit a self-supporting argument for a counter-inductive rule which was precisely parallel to the self-supporting argument for the standard inductive rule. In (Salmon [1967a], p. 14), I showed that parallel self-supporting arguments could be provided for *modus ponens* and for the fallacy of affirming the consequent. The vicious circularity of such arguments is revealed by the fact that they can be devised even for manifestly unsatisfactory rules. That which equally supports every rule supports no rule at all.

This is precisely my objection to arbitrary rules. If any case can be made for the adoption of one arbitrary rule, then an equally good case can be made for adopting another, as the foregoing discussion has shown. For any conclusion which is licensed by one arbitrary rule on the basis of a given body of evidence, a different and conflicting conclusion can be drawn on the basis of the same evidence by invoking a different arbitrary rule. Indeed, any conclusion which is logically compatible with the evidence is sanctioned by some rule or other. These are sufficient grounds for condemning arbitarariness in our rules of inductive inference.

Strawson and many others have claimed that the use of standard inductive rules is rational because that is part of what we ordinarily mean by the word "rational". This approach to the concept of rationality has always struck me as extraordinarily inadequate, and it still does (Salmon [1957a]; [1965a]; [1967a], Chap. II, Section 7). I have sometimes said that I cannot understand any concept of inductive rationality which does not have some direct connection with frequency of success in making posits or predictions (Salmon [1965b]). As a result of Clendinnen's arguments, I now think that my former view was too extreme. Even if we cannot show that adoption of arbitrary rules of inference will yield smaller success ratios than does the standard rule, we can still make a good case that their adoption would be irrational. If this conclusion is correct, it has an *extremely important* bearing upon the problem of the justification of induction. This is what I see as the basic significance of Clendinnen's discussion. Although he and I differ on details, I believe we are in wholehearted agreement on fundamental principles.

It is reasonable to expect that, when the history of twentieth century philosophy is written, one important chapter will be devoted to inductive logic and probability theory, and that Carnap's contributions will occupy a primary place in that chapter (see Salmon [1967b]). Nevertheless, in spite of my enormous admiration for Carnap's work, I am convinced that he failed adequately to come to terms with what *I take to be* Hume's problem of the justification of induction (Salmon [1967a], Chap. IV, Section 3).

David Stove has offered a careful and penetrating criticism of my claim that the Carnapian approach to inductive logic does not resolve Hume's problem of the justification of induction. The critique is both systematic and historical, with a bit of psychoanalysis thrown in for good measure. In order to sort out the main issues, a few preliminary points of clarification should be made.

First, Stove uses the term "inductive argument" in a rather narrow sense to designate "an argument from observed to unobserved instances of some empirical predicates" (Section II), while I use it to refer to non-demonstrative inferences quite generally (Salmon [1967a], p. 8). In the present context this difference has little import, for if we are discussing inferences conforming to Reichenbach's rule of induction by enumeration or those represented by Carnap's confirmation functions, these inferences qualify as inductive in both senses. However, it seems to me, Hume's concern was not restricted to inductive inferences in the narrower sense, for he was raising questions about "reasonings concerning matters of fact and existence" and "the nature of that evidence which assures us of any real existence and matters of fact beyond the present testimony of our sense or the records of our memory". Thus, I take it that Hume's inductive skepticism applies to inductive inference construed in the broader sense.

Second, I agree heartily with Stove on the importance of distinguishing between inductive fallibilism and inductive skepticism. Hume was not the first philosopher to realize that inferences from the observed to the unobserved do not carry necessity; the ancient skeptics had long since successfully established the fallibilistic thesis, though perhaps it had sometimes been forgotten in the interim. Hume's revolutionary thesis is inductive skepticism, which may, if we are careful about terminology, be paraphrased in just the way Stove suggests, namely, "that the conclusions of inductive arguments, in relation to their premises, *are not probable*" (Section II). Having agreed to this formulation, I must insist that we exercise extreme care in our use of the term "probable".

Third, I must reject with a smile Stove's psychological explanation of philosophers' preoccupation with the problem which Stove considers to have been incorrectly attributed to Hume by contemporary philosophers of science. (I get the distinct impression that I am one of the chief offenders.) I do not think I have ever confused inductive fallibilism with inductive skepticism; indeed, I have been severely critical of those philosophers who believe that Hume's problem of induction vanishes once we are able to give up "the quest for certainty". I have never been shocked out of my mind on account of the futility of that quest, nor do I consider my "philosophic fears" to be "morbidly nervous". *That* is not a source of *my* preoccupation with the problem of justifying inductive reasoning. Fallibilism is, I take it, a presupposition of the discussion of Hume's problem of induction — whatever that problem turns out to be. I have not a shadow of doubt that it is reasonable to have high degrees of belief in propositions strongly supported (though to a degree falling short of deductive certainty) by evidence. The problem concerns the nature of evidential support (see Salmon [1965a]).

Fourth, I think it is important to point out that, although isolated statements may give a different impression, I did not raise Stove's question (Q2) (Section I) in my critique of Carnap, but rather, a question much more closely akin to his (Q2′) (Section III). Thus, I readily acknowledge that (Q2) involves a *suggestio falsi*, and agree, by and large, with Stove's repudiation of it. However, the appropriate interpretation of this question (Q2′) does not emerge until Stove's suggestion (Section IV) that the word "probable" in (Q2) (or, presumably, the word "probability" in (Q2′)) is enclosed in invisible scare quotes. It is not until this last section of Stove's essay that any major systematic, as opposed to historical, issue arises.

Let me now attempt to exhibit what I take to be the fundamental issue. Under the interpretation of Stove's Section IV, (Q2′) takes the following form (where the scare quotes are made visible, and an inessential parenthetical remark is deleted):

(Q2″)   Why should one have, in a proposition H which has 'probability' $= x/y$ in relation to one's total evidence E, a degree of belief which is a fraction x/y of one's degree of belief in E?

It should be mentioned in passing that, as Richard Jeffrey has shown, a degree of belief equal to x/y *of one's degree of belief in E* is appropriate only if one's degree of belief in E is unity. In this discussion, we may go along with the assumption that we always have full confidence in our evidence statements, for that supposition is built into Carnap's systems of inductive logic. The crucial question is how to construe the term "probability". Stove says that Carnap gave us the answer in 1950, presumably, that "probability" means Carnap's probability$_1$ which was furnished by the confirmation function c^*. Let us make the appropriate substitution in (Q2″) and do a bit of rewording:

(Q2‴)   Given that $c^*(H, E) = x/y$, and that E constitutes one's total evidence (in which one's degree of belief $= 1$), why should one have a degree of belief equal to x/y in H?

Unlike Stove's question (Q2′), this question, to my mind, gives no appearance of being a candidate for repudiation.

If we agree with Stove's remark, "The very thing which assessments of probability do . . . is to characterize certain *degrees* of belief as rational . . . " (Section III), then it seems to me that we cannot help raising the question, which seems to have occupied much of Carnap's attention after 1950:

(Q3)   Can c^* legitimately be identified with probability in the sense of "rational degree of belief"?

I think this is a serious question, and I believe Stove does as well. He thinks the answer is affirmative, and that this affirmative answer can be justified. He supports this claim by showing how a fragment of inductive logic can be derived from "incontestable" logical principles. I intend to question the significance of this fragment of inductive logic, and to contest the symmetry principle to which he appeals.

Before 1950, some philosophers (e.g., Wittgenstein) seemed to advocate a confirmation function $c^\dagger$, but Carnap rejected this function because it precludes "learning from experience". He advocated, instead, the confirmation function c^*, which does not have this defect. Both $c^\dagger$ and c^* are symmetric; hence, both satisfy the symmetry principle to which Stove appeals. If we want to make even a singular predictive inference – i.e., an inference from an observed sample to an unobserved member of the same population – we must decide between these two functions, and that will require further assumptions. Moreover, c^* has other draw-backs, including the fact that universal generalizations can never have degrees of confirmation greater than zero on any finite body of observational evidence (Carnap [1950], Section 110). In his [1952], Carnap had given up hope of justifying the selection of a unique inductive method, and he gave us a continuum of methods from which to choose. In (Carnap [1963]), he offered 14 axioms upon which the selection of inductive methods is to be based. These axioms, whose ultimate justification is "inductive intuition", are neither simple, nor seemingly innocuous, nor incontestable. Axiom 13, "the axiom of convergence", has the same import as Reichenbach's pragmatic justification of asymptotic rules. Axiom 12, "the axiom of instantial relevance", amounts to the assertion – sharply questioned by Hume – that the past *is* a guide to the future. Russell's famous remark about the method of postulation is especially apt with regard to such axioms: this approach does have all the advantages of theft over honest toil. I believe that Carnap's inductive intuition is not an adequate basis for inductive logic; *some* heavy labor seems indispensable.

Stove makes no pretence of justifying a full-blown inductive logic; he provides, in effect, that portion shared by all of the symmetric confirmation functions (Carnap [1950], Chap. VIII). Regarding the principle of symmetry with respect to individual constants, he says that it "is so indispensably necessary even for deductive logic that no one could afford to deny [it]" (Section IV). It is not altogether clear to me in what sense this principle is indispensable to deductive logic, but perhaps he means something like this: Since "All men are mortal and Socrates is a man; therefore, Socrates is mortal" is a valid argument, everyone must admit that "All men are mortal and Plato is a man; therefore, Plato is mortal" must also be valid. It should be noted in addition that the validity of "All men are featherless and Socrates is a man; therefore, Socrates is featherless" is assured by comparison with the first of the two foregoing arguments. Thus, one might say, a principle of symmetry with respect to predicates is also a basic characteristic of deductive logic. It is, however, a general feature of Carnapian inductive logics to violate this unrestricted principle of symmetry with respect to predicates. The extension of symmetry principles from deductive to inductive logic must therefore be handled with great caution; substitutions which preserve deductive validity cannot be counted upon to preserve probability relations.

The principle of symmetry with respect to individual constants constitutes a special case of the principle of indifference. Carnap was fully aware that unbridled use of the principle of indifference leads to logical inconsistency, but he looked for a valid core of the traditional principle which could safely be incorporated into inductive logic. Carnap regarded symmetry with respect to individual constants in this light. Given, however, the dubious status of the traditional principle of indifference, it seems to me unwise not to subject its special cases to critical scrutiny.

If we are to develop a concept of logical or inductive probability, along the lines advocated by Carnap, we might follow his general strategy of laying down the initial requirement (axioms 1–5 in Carnap [1963]), that confirmation functions must satisfy the mathematical calculus of probability. We might then insist (ibid., axiom 6) upon a regularity condition which, in finite languages, permits the probability of h on e to be 1 only if e logically entails h. This principle – which Carnap invoked, in effect, to reject the "straight rule" (Carnap [1952], Section 14) – strikes me as rather questionable. But if we grant it, then the next step ([1963], axiom 7) might be the principle of invariance with respect to finite permutations of individuals – the symmetry principle invoked by Stove. All of this is, however, conditional upon our willingness to embark upon the project of constructing an inductive logic based upon a concept of logical probability, an undertaking whose advisability I attempted to call into serious question in my [1969c]. I must plead guilty to Stove's charge (Section II) that "probabilities leave

[me] cold", if, by "probabilities", he means *logical* probabilities. Other concepts of probability — especially frequencies (Carnap's probability$_2$) — strike me as interesting, robust, and powerful. Carnap's wife once remarked, in Carnap's presence, that at home they referred to probability$_1$ as "little Rudi" and probability$_2$ as "big Rudi". The reason Carnap devoted so much attention to probability$_1$ and so little, comparatively, to probability$_2$ was that little Rudi was the one who needed help. Big Rudi could take care of himself. My own opinion is that little Rudi never did become a viable independent entity, and Mellor seems to hold a similar view.

On the historical side, Stove (Section II, esp. note 8) challenges me to document — preferably by citations from the *Treatise* — the claim that the problem I attribute to Hume is, indeed a problem Hume raised. The most natural place to look is Book I, Part IV, Section VII, the summary of Book I. There Hume expresses his skepticism (*not* just fallibilism) in eloquent terms. " ... [T]he understanding, when it acts alone, and according to its most general principles, entirely subverts itself, and leaves not the lowest degree of evidence in any proposition, either in philosophy or in common life" (Selby-Bigge [1888], pp. 267–268). "The *intense* view of these manifold contradictions and imperfections in human reason has so wrought upon me, and heated my brain, that I am ready to reject all belief and reasoning, and can look upon no opinion even as more probable or likely than another" (ibid., pp. 268–269). One could demand more detailed documentation, of course, and I could supply it. However, since I agree with Stove's formation of Hume's inductive skepticism (as I remarked above), it hardly seems necessary to devote more space to that exercise.

The problem Hume raises can be put in the following way. Let us grant that some propositions (regarding, say, relations of ideas) can be known with certainty. Let us grant that some propositions (regarding, say, weather conditions in the remote future) are totally unsupported by available evidence. Is it possible — and if so, how? — to secure an intermediate status, somewhere between *supported with certainty by the evidence* and *totally unsupported by evidence*, for any other propositions concerning unobserved matters of fact? If it could be shown that a proposition is strongly supported by evidence, but to a degree falling somewhat short of certainty, then we could legitimately say that it is probable (in Hume's sense of the word) on the basis of the pertinent evidence. Hume says, in the above-quoted passages, and in many other places as well, that he does not see how to render any such proposition probable on the basis of empirical evidence.

One might have said to Hume, "Why not assign probabilities to propositions in the way people usually do?" Stove does seem to admit that it makes sense to ask "whether what *passes with us* for probable really is so" (Section IV), and he seems to say that an affirmative answer can legitimately be given if these probability assignments agree with Carnap's inductive logic. Unfortunately, no one had provided that answer prior to 1776, so we cannot know for sure what Hume would have said in response. But I have said what I think a philosopher who has appreciated Hume's writings *should* ask, namely, "Why should one's degree of belief in a proposition correspond to its probability$_1$ with respect to the total available evidence?" This question is essentially the same as Stove's (Q2'), except that the subscript "1" has been attached to the word "probability", signifying that we are now talking about Carnap's concept of inductive probability rather than some ill-specified ordinary sense of the term. This question is tantamount to question (Q2''') which I set out above. As I tried to show, this question naturally leads to (Q3). I have already stated my reasons for denying that an affirmative answer to (Q3) has been provided.

As long as we have a plethora of confirmation functions – i.e., definitions of inductive probability – to consider, it certainly makes good sense to ask, with respect to any particular inductive probability function, why we should use probability in *that* sense to determine our degrees of belief. If it is an historical injustice to claim that this is what Hume's problem about induction comes down to after we have worked through the putative answer Carnap offers, I apologize. I have no wish to claim Humean fatherhood for a bastard I have sired. But if Hume is not the father, he is surely the grandfather.

In analyzing various modern approaches to induction, I have tried, in each case, to see how it copes with the basic problem Hume raised concerning inference from the observed to the unobserved (Salmon [1967a], Chap. V). I have just indicated how it comes out after we work through the Carnapian systems of inductive probability. But I never meant to suggest that this is the only approach which runs into difficulty with the basic Humean problem. If we reject the use of inductive probabilities, and try to establish an inductive logic based upon statistical probabilities, the same fundamental problem crops up. It does not sound quite the same, but it is there nevertheless. In (Salmon [1967a], Chap. V) I tried to spell out the parallels. The attempt to justify a rule for positing long-run frequencies constitutes the manner of dealing with the very same problem within the frequency approach. In attempting to see how these diverse approaches cope with Hume's problem, I am not attempting to *do* history, I am *using* history. But I do not think this is a misuse of history.

Similar remarks could be made, by way of analogy, about Zeno's paradox of the flying arrow. The problem arises, it might be said, because Zeno did not understand the concept of instantaneous velocity – a concept which was not clarified until the infinitesimal calculus was available. When we have that concept, however, we can make the distinction – inaccessible to Zeno – between instantaneous rest and instantaneous motion. Thus the paradox is resolved, it may be said. But this conclusion is not justified, for when we examine the concept of instantaneous velocity, as it is defined in the calculus, we find that it is the limit of a sequence of average velocities taken over decreasing intervals. To make sense of instantaneous velocity, it turns out, we must have recourse to motion over an interval, which is precisely the concept which gave rise to the paradox of the flying arrow in the first place (Salmon [1975a], pp. 38–42). To admit a superficial answer to a deep question is to do an injustice to history, I believe, even if the historical philosopher did not have available the later conceptual apparatus upon which the superficial answer is based. In spite of my enormous admiration for Carnap's work on inductive logic, I do not think he provided a penetrating answer to Hume's inductive skepticism.

Among the most significant developments in the theory of probability in the 1950's were the emergence of the theory of personal probabilities (due mainly to L. J. Savage) and the propensity theory (initiated by Karl Popper). D. H. Mellor's *The Matter of Chance* [1971] brought the two together in an extremely interesting way. An invitation to do a lengthy discussion-review (Salmon [1979f]) of that book afforded me an opportunity to treat these developments comprehensively. Concerning the propensity theory, I tried to make three points: (1) That the defects Popper found in the frequency interpretation, which led him to propose a new interpretation, were not present in Reichenbach's frequency theory; indeed, the solutions Popper offers by way of his propensity theory do not differ basically from those which Reichenbach had already offered in the context of his frequency theory. (2) That Popper's claims about the ability of the propensity interpretation to solve the philosophical problems of quantum mechanics are patently unfounded. (3) That, while propensity may be an interesting concept, it does not furnish an interpretation of the probability calculus. My argument on this point is, I must hasten to remark, altogether different from that of Suppes, which Mellor answers in 'Chance and Degrees of Belief'; moreover, it applies only to versions of the propensity theory which Mellor explicitly repudiates (Section 3), and does not affect his theory at all.

In my review, I did level a number of serious criticisms at Mellor's theory, and I am grateful to him for taking them in excellent spirit, and for responding with a clear and penetrating essay in the epistemology of probability. As it turns out, we see eye to eye on a number of important points. For example, I am in complete agreement with his assessment (Section II) of the shortcomings of purely subjective probabilities in the context of Bayesian decision theory, and also with his view that Carnapian inductive probabilities are insufficient to fill the gap. "The work chance facts do is justifying degrees of belief." " . . . [I]nductive probabilities are not enough to make sense of the prescriptions of Bayesian decision theory. The work objective probability has to do there can only be done by chance."

In Section 7, Mellor discusses his use of the law of large numbers to establish a relationship between chances and degrees of belief. I had challenged his use of that law on several scores. First, I pointed to the fact that this law has as an hypothesis a condition of independence, and I suggested that we have no right to appeal to it unless we know that this condition is fulfilled. Although Mellor attempted to deal with this issue by stipulating that the chance distribution is the same on every trial ([1971], p. 161), it still seems to me that we have a serious problem in assuring that it is, in fact, satisfied. However, I shall not pursue this question farther here. Second, I raised objections against his argument about "breaking even", but on this point it appears that I simply misconstrued his argument. When he spoke of a fixed stake, I thought he was referring to the amount staked on each bet. If we then consider what happens on more bets, we have to put up a larger sum of money. His intent, as he later pointed out in conversation, was that the total stake is to be fixed, so that consideration of a larger number of bets entails a proportionate diminishing of the amount staked on each individual bet (cf. ibid.). When the situation is construed in that way, his conclusion about breaking even follows without difficulty. In any case, as he says at the end of Section 8, we completely agree that a coherent betting quotient which equals the chance p enjoys a privileged position among CBQ's.

Third, and most importantly, I found myself uneasy about Mellor's treatment of the second level probability p^* which occurs in the law of large numbers. As Mellor remarks, "This very high degree p^* of belief in the propositions F looks as if it needs another chance to justify it; and if that were so, there would be an arguably vicious circle. But p^* does not need a chance to justify it . . . " Let us see what, precisely, is the status of this probability p^*. Mellor tells us (ibid., p. 162) that it is a CBQ, which is to say that, as a degree of belief, it cannot be faulted for leading to a violation of the mathematical rules of probability. But as Mellor says explicitly in 'Chance and Degrees of Belief", this minimal brand of coherence is not sufficient to guarantee that a degree of belief is *reasonable*. He wants to maintain, nevertheless, that it is *reasonable* to have a large degree of belief in the statement that the finite frequency in a sufficiently large number of trials will closely match the chance on each of the trials which make up the aggregate. Mellor acknowledges explicitly, both in the book and in the present essay, that any attempt to justify this degree of belief p^* on the basis of chance would threaten either circularity or vicious regress. So that avenue is not open. I would be happy enough if he were to offer a frequency justification — that one who uses p^* in this way will usually be right — but one of his major concerns is to eschew such appeals to frequencies. Hence, this avenue, as well, is closed.

What Mellor appears to do is to argue that the law of large numbers furnishes us with a sequence of CBQ's, associated with larger and larger values of N, which converges to one — that is, a sequence of partial beliefs which converge toward full belief. That much is true. He then seems to claim that, because they thus converge toward full belief, they ipso facto qualify as *rational* degrees of belief (ibid., pp. 62–64). This must surely be a non-sequitur. Suppose I believe to the degree $1 - (\frac{1}{2})^N$ that a werewolf will appear at the next full moon on Bodmin Moor (Cornwall), where N is the number of consecutive full moons which have appeared there unobscured by clouds immediately prior to the full moon in question. If a sequence of N such full moons occurs at Bodmin Moor, my degree of belief approaches full belief, as N increases, that a werewolf will appear; a sequence of 13 such lunar months (about one calendar year) would produce a degree of belief just slightly less than 0.9999. I do not see how such beliefs can be shown to be incoherent; they do seem to qualify as CBQ's. However, the fact that they converge toward unity does not make them reasonable. I cannot find any argument in Mellor's discussion to show that his p^*'s, which admittedly converge to 1, constitute reasonable CBQ's. Perhaps I am simply being obtuse in failing to grasp a crucial point in his treatment of this issue; the reader must decide. Either way this problem does lie at the crux of his approach to chance.

I am not prepared to give up on frequencies quite yet. If we are dealing with a sequence of independent trials, and if the first level probability (chance) p represents a limiting frequency, then the law of large numbers can legitimately be interpreted to say that the second level probability p^* also stands for a limiting frequency. This means that one who uses p^* as a betting quotient will be right most of the time in the long run. Although this use of frequencies is not unproblematic, it does seem to do the work required of objective chance. Moreover, I think it is possible to overcome Mellor's basic objection to frequencies. If a rule of induction, of the sort discussed at length above, can be taken to be part of our intellectual apparatus, then we can say that a finite frequency in an observed sample *can* cause us to posit a value for the limiting frequency. *These* observed frequencies *are* "causally available in advance".

Robert McLaughlin's essay, 'Invention and Appraisal', contains an admirable discussion of the discovery vs. justification issue which has periodically erupted in the literature over the past several decades. It is valuable, first, because of its initial terminological proposal. For reasons which he clearly states, the terms "invention/appraisal" constitute a significant improvement over the traditional "discovery/justification". McLaughlin offers enough evidence of purely verbal confusion to provide ample motivation for this terminological reform.

Second, in his treatment of *advancement* and *enhancement* arguments, McLaughlin exhibits the crucial role played by prior probabilities – plausibility considerations – both in the context of invention and in the context of appraisal. It seems to me that a good deal of misguided argument has infected discovery/justification controversies in the past precisely because of a failure of the parties to appreciate fully the Bayesian aspects of scientific appraisal (justification). In his treatment of the role of prior probabilities in advancement as well as enhancement, McLaughlin makes the strongest case, it seems to me, for a logic of invention (discovery). I would be somewhat inclined to call it a "heuristic" rather than a "logic", but whichever term is adopted, we must surely admit that invention (discovery) has certain rational aspects which can be philosophically explicated. That is the fundamental philosophical point. I should mention, incidentally, that I find McLaughlin's argument for a variety of plausibility considerations – including especially analogy – convincing, as against Clendinnen's view that all plausibility considerations can be reduced to appeals to simplicity. Analogical arguments constitute an important and widely used class of plausibility considerations. Let me add just one more example to those given by McLaughlin. Archaeologists (prehistorians) often appeal to ethnographic analogies – that is, studies of extant cultures which can be supposed to have important similarities to prehistoric cultures with which they are concerned. Such analogies are well-suited to play a role both in invention and in appraisal of archaeological hypotheses.

Third, McLaughlin's examination of the received view of Popper and

Reichenbach, the inventionist view of Achinstein and Hanson, and the icono-
clast view of Kuhn and Feyerabend is the most fair and balanced treatment
I have seen. His discussion, unlike most others, exhibits a full understanding
of the motivations, strengths, and weaknesses of all three.

One of the most persistent and significant issues in philosophy of science
in recent years has been the controversy between those who might roughly
be characterized as logical empiricists and those who adopt an historical
approach. A good deal of the debate has hinged, directly or indirectly, upon
the logical status of scientific invention (discovery), and a fair amount of
confusion has been generated out of various misunderstandings regarding
the role of plausibility considerations. What an historian may identify as
a non-rational factor entering into the acceptance or rejection of a scientific
theory may, in fact, turn out to be a plausibility consideration which plays
a legitimate role in either a rational advancement argument, or a rational enhancement argument, or both. If the ideas embodied in McLaughlin's essay
could be brought to bear upon some of the controversies between these two
major schools of thought, a great deal of needless conflict could, I think,
be alleviated. This point will be exemplified in the next section.

SPACE AND TIME

There is much with which to agree in the illuminating historical discussion of
Einsteinian simultaneity by Saunders and Norton. They are quite right, I
believe, in their remarks about the ontological status of fields, and about
the importance of the fact that light is a "first signal". I feel somewhat
uneasy about their assertion that "space will not allow signals to propagate
faster than a certain maximum speed", but I would be quite comfortable
with the claim that a maximal speed of propagation characterizes the electromagnetic field itself. Looking at this matter from the standpoint of Minkowski
spacetime geometry, we see that the most significant feature of light signals
is that they furnish the light cone, which can appropriately be regarded as the
cone of causal connectability. For those who (like Reichenbach, Grünbaum,
and myself) desire to adopt a causal theory of space and time, it is not that
space (or spacetime) prevents certain kinds of behaviour on the part of
signals; rather the behaviour of light signals as first signals provides the
fundamental spacetime structure. This approach is nicely elaborated by
John Winnie [1977a].

The main purpose of the Saunders/Norton essay is to examine a thesis
about the conventionality of simultaneity which has been advocated chiefly

by Reichenbach and Grünbaum, but also by others (including myself). They acknowledge (in the concluding paragraphs of Section 2) that this thesis is irrefutable on strictly empirical grounds, but they argue that these grounds are too restricted to settle the issue of conventionality. "The conventionalist thesis is bound by a narrow epistemology", they maintain. "The problem with this epistemology manifests itself in the failure of the conventionalists to allow that there can be factors involved in the choice of theories that are non-empirical in character, factors that appear purely on a theoretical level." To set the record straight, they propose to examine the historical question of "why Einstein chose standard signal synchrony and not another". They claim that this examination will show that the contemporary conventionalism I have espoused goes beyond Einstein's conception of definition in his classic 1905 paper on special relativity. I should like to comment upon both their philosophical thesis and their historical claim.

The cornerstone of Reichenbach's conventionality thesis is his distinction ([1938], Section 42) between the "context of discovery" and the "context of justification" – better denoted by McLaughlin's terms, "invention" and "appraisal". If we recognize this invention/appraisal distinction, then it is clear that *choice* – either theory choice or choice of a synchrony definition – involves elements from both the context of appraisal and the context of invention as well. The repeated references by Saunders and Norton to heuristics clearly reveal the fact that the context of invention is deeply involved in their discussion. What Saunders and Norton say about heuristics is interesting and valuable, but it has no bearing whatever upon the problem Reichenbach and other conventionalists are addressing. Reichenbach explicitly emphasized his concern with the rational reconstruction of the special theory of relativity; his considerations fall within his context of justification, while those relating to heuristics belong to his context of discovery. Thus, one could concede all of the points made by Saunders and Norton regarding the heuristics of theory choice, and still maintain Reichenbach's conventionality thesis with respect to simultaneity. Since Reichenbach was fully aware of the heuristic elements, it is simply not correct to say that conventionalists of his persuasion do not "allow that there can be factors involved in the choice of theories that are non-empirical in character". To distinguish between invention and appraisal is not to deny the existence of the context of invention, or to deny that it has any importance.

Let us turn to the context of appraisal. Here I think it might be helpful to distinguish two sorts of appraisal – cognitive appraisal and pragmatic appraisal. If it is said that scientists prefer simple theories, this might, as

Reichenbach insisted, mean either of two things, for as I explained in the preceding section, he distinguished two types of simplicity, descriptive and inductive. Descriptive simplicity, it will be recalled, applies in cases where a choice must be made among theories which are factually equivalent to one another – among equivalent descriptions. The alternative theories say the same things, and the choice is made on the basis of economic or aesthetic considerations. Since the theories do not differ in content, there can be no cognitive basis for preferring one to the others. None can be shown to be more or less probable than any of the others. There is, however, a pragmatic basis for choice, because the simplest description is more efficient and, perhaps, more beautiful. Reichenbach maintained that Einstein's choice of standard signal synchrony ($\epsilon = \frac{1}{2}$) *can be justified* on the basis of pragmatic considerations inasmuch as it leads to descriptively simpler theories than other definitions of distant simultaneity would yield. Such a choice results from a pragmatic appraisal. Conventionalists do not deny the enormous importance of appraisals of this kind, but they do insist that the alternative descriptions do not differ in factual content.

Saunders and Norton seem to maintain that, even if we accept the invention/appraisal distinction, and the pragmatic/cognitive distinction within the context of appraisal, there is still more to Einstein's choice of standard signal synchrony than I have acknowledged in the heuristic and the pragmatic-appraisal aspects. They are claiming, as I understand them, that there are non-empirical aspects of theory choice which bear upon the cognitive appraisal of theories. They attempt to argue that the various theories which result from different choices of ϵ differ from one another in theoretical content.

Although I do not agree that there are non-empirical considerations which properly affect the cognitive appraisal of theories, I do think it is possible – indeed, essential – to distinguish two types of empirical evidence upon which the evaluation of theories rests. If one accepts, as I do, a Bayesian account of confirmation, then it turns out that two kinds of probabilities – prior probabilities and likelihoods – are required to establish the posterior probability of a hypothesis (Salmon [1967a], Chap. VII). I believe that both kinds of evidence are needed, and that both are empirical and objective. Now it seems clear to me that simplicity and spatial isotropy are factors which might well bear upon the prior probability, or plausibility, of a theory like special relativity. Such considerations, which I take to be based upon empirical experience with other physical theories, are entirely admissible in the cognitive appraisal of theories, but they are quite different from direct experimental tests of the theories. It might not be too misleading to say that the plausibility considerations constitute broadly empirical evidence, while the results of direct tests constitute narrowly empirical evidence. However, it is a misconception to suppose that the plausibility considerations are non-empirical.

With these general ideas in mind, let us now turn more specifically to the question of why Einstein chose the particular definition of simultaneity which he did. In addressing this question, Saunders and Norton place considerable emphasis upon two statements which occur near the beginning of Einstein's 1905 paper. The first occurs in the introductory paragraphs before the beginning of Section 1, where Einstein writes, "We will ... also introduce another postulate ... namely, that light is always propagated in empty space with a definite velocity c which is independent of the state of motion of the emitting body" (p. 38). In this passage, Einstein is telling us what he is going to do; he is not setting up his postulates as yet. He then proceeds to explain the need for a definition of distant simultaneity, and to present his famous definition in terms of standard signal synchrony. Given that definition, statements about one-way velocities are now meaningful. He concludes Section 1 with the remark, "It is essential to have time *defined* by means of stationary clocks in the stationary system, and the time now *defined* being appropriate to the stationary system we call it 'the time of the stationary system'" (p. 40, my italics).

With the definition of simultaneity available, Einstein officially introduces his two postulates in Section 2. The first postulate is the principle of relativity; the second is the constancy of the speed of light. The precise statement of this second postulate, which had been anticipated in the opening paragraphs, is as follows:

Any ray of light moves in the "stationary" system of coordinates with the determined velocity c, whether the ray be emitted by a stationary or by a moving body. Hence

$$\text{velocity} = \frac{\text{light path}}{\text{time interval}}$$

where time interval is to be taken in the sense of the definition in Section 1 (p. 41, my italics).

Since this postulate plays a crucial role in the argument of Saunders and Norton (Section 3) that " ... Einstein's choice of $\epsilon = \frac{1}{2}$ was not based on questions of conceptual or manipulative convenience, as the conventionalist account would suggest, but on the basis of prior theoretical commitments", it is essential to notice the explicit reference back to his definition. This prior theoretical commitment apparently cannot even be articulated without the definition of distant simultaneity. These quotations seem to me to support unequivocally the claim that Einstein regarded his definition as a definition. If Einstein had regarded the one-way speed of light as an objective fact, I do not see why he would have qualified his light postulate with the second sentence – unmentioned by Saunders and Norton – which refers back to the *definition* of distant simultaneity.

Immediately following his definition of distant simultaneity, Einstein states that, "in agreement with experience", the average round-trip speed of light is taken to have the constant value c. One way to look at the Michelson-Morley experiment is to say that it establishes that this average round-trip speed is the same regardless of the orientation of that path in space. This conclusion can be taken as giving straightforward experimental meaning to the thesis that space is isotropic. The isotropy of space is an important physical thesis, and it may have figured significantly in Einstein's thought about special relativity, but it can be formulated – as I have just done – in terms of average round-trip speeds. Let me call this "two-way isotropy". Two-way isotropy well might qualify as a cogent plausibility consideration.

Given *that* sort of isotropy, and given the fact that light is a first signal, we see that we can extend the isotropy of space, *by definition*, to the principle that the one-way speed of light is constant in all directions (one-way isotropy). This latter formulation does not, as far as I can see, involve any physical assertion about the factual equality of the one-way speeds without first providing a *definition* of distant simultaneity. As Einstein said so very clearly, to assign a definite value to the speed of an object traveling from A to B, we must be able to compare A-times with B-times. That comparison demands a convention.

Saunders and Norton are right, I think, in maintaining that one of Einstein's fundamental aims in developing special relativity was to find a theory which would embody the standard Lorentz transformations. He chose the definition $\epsilon = \frac{1}{2}$, and bingo! – out came the Lorentz transformations. As far as I know, Einstein never entertained the notion of adopting any other definition of simultaneity, but that does not seem to me to imply that he thought his definition was anything other than a definition. A straightforward interpretation is that the particular definition he chose did exactly what he wanted it to do – there was no need to hunt around for other definitions. The crucial physical fact is that the world admits of the description – formulated in terms of the standard definition of simultaneity and the Lorentz transformations

– given by the special theory of relativity. Even if Winnie's ϵ-Lorentz formulations (Winnie [1970]) are physically admissible (allowing for choices of ϵ other than $\frac{1}{2}$), that does not undermine the claim that the standard formulation in terms of standard signal synchrony provides a physically adequate description. Likewise, one can argue, the crucial physical fact about the isotropy of space is that the one-way isotropic description is physically admissible. It does not seem to me that any fact of empirical or theoretical importance is left out when the matter is presented in these terms.

There is one further historical point which I should like to clarify. Some of the early logical positivists were phenomenalists, and Reichenbach is often called a positivist, but as a matter of fact he never was one. He regarded his major epistemological treatise, *Experience and Prediction* [1938], as a refutation of logical positivism, and one of the chief points in his argument against positivism is his rejection of phenomenalism. Reichenbach was a physicalist; he held that statements about ordinary physical objects are directly supported by perceptual evidence. Sense data were, on his view, theoretical entities in psychology. Moreover, he rejected positivistic instrumentalism. He was a theoretical realist; he held that it is possible to have indirect inductive evidence for the existence of the kinds of unobservable objects to which scientific theories refer. Furthermore, Grünbaum and I are *not* motivated by phenomenalistic doctrine, or by any reluctance to admit theoretical considerations. All of us are, however, concerned with the evidential support which can be summoned for theoretical assertions.

My conclusion is that Saunders and Norton have told us a good deal about the heuristic value of standard signal synchrony, and about its bearing upon the pragmatic appraisal of theories, but they have not undercut the thesis of conventionality of simultaneity when it is understood – as intended by Reichenbach, Grünbaum, and me – as a thesis regarding the cognitive appraisal of theories. I cannot resist remarking that Kuhn's influential repudiation [1962] of the discovery/justification (or invention/appraisal) distinction has had adverse effects upon discussions of such issues as the one at hand, and I cannot resist expressing the hope that attention to McLaughlin's treatment of the distinction will help undo some of the damage.

The Saunders/Norton discussion of simultaneity is conducted within the context of the Einsteinian approach utilizing rods, clocks, and frames of reference; Nerlich staunchly advocates an approach which employs Minkowski spacetime geometry. I am thoroughly convinced of the value of adopting this latter approach to special relativity, especially in view of the beautiful results in Winnie [1977a] and Malament [1977]. I am not sure it would be advisable to abandon altogether the Einstein rods-and-clocks approach, but I do believe that the Minkowski spacetime approach furnishes important insights which are not readily apparent in the other approach. It is always worthwhile to look at important physical theories from different viewpoints.

There is, however, one fundamental point of disagreement between Nerlich on the one hand and Winnie and Malament on the other. Winnie – elaborating ideas which were originated by A. A. Robb in the early years of relativity theory, but which remained almost totally unknown – develops a *causal theory* of spacetime. In that theory, the fundamental physical distinction among the three kinds of separation – space-like, time-like, and light-like – is founded upon the relation of causal connectability. Malament adopts a similarly causal approach. Nerlich, in contrast, rejects the causal theory (see note 13, p. 153). This rejection stems from his claim (see, for example, the closing paragraphs of Section 3) that it is an error to suppose that special relativity confers any special status upon time-like lines. On this point I disagree most emphatically.

My main reasons for taking issue with Nerlich regarding causality are sketched in 'Comets, Pollen, and Dreams'. If we look at the world from the standpoint of 4-dimensional Minkowski spacetime, we can identify certain spatio-temporally continuous lines which are paths of various kinds of physical processes – e.g., stationary or moving material particles, light pulses, and moving shadows. As I tried to show, we can apply the mark method to distinguish which of these processes qualify as genuine causal processes and which must be classed as pseudo-processes. Extensive empirical investigation establishes the fact that the worldlines of causal processes are invariably time-like or light-like, but never space-like. Pseudo-processes, in contrast, may have space-like, light-like, or time-like worldlines.

In 'An 'At-At' Theory of Causal Influence' [1977f], I attempted to argue that causal processes constitute the physical means by which causal influence is transmitted from events in one region of spacetime to other regions. Such processes can also be said to transmit energy, information, structure and order. They *constitute* the physical connections among the events which make up our world. This seems to me to be ample reason to assert that time-like and light-like lines enjoy a special status which is not shared by space-like lines.

Certain kinds of causal processes – such as light propagation – define a cone which encompasses all possible trajectories of causal processes. The surface of this cone contains the null geodesics which are the trajectories of all possible first signals. This fact gives those kinds of processes – the first signals – a crucial role in determining the causal structure of the physical world. As Winnie [1977a] shows with admirable clarity, the entire structure of Minkowski spacetime can be defined on the basis of these relations of causal connectability. A physical theory which ignored the role of causal processes and causal connections would be, I submit, patently inadequate. Special relativity is, as Nerlich claims, a physics of fields as well as particles, but it is essential to remember that disturbances are propagated through fields at finite velocities. There is no direct causal connection between spatially separated parts of the field at the same time, but different temporal slices of a particle are causally connected by virtue of being parts of a single causal process.

Nerlich's essay raises certain profound and interesting questions about the concept of a frame of reference. Although this concept was conspicuously present in Einstein's 1905 paper – and can be found in the great majority of subsequent presentations of special relativity – it does involve serious ambiguities, as Nerlich demonstrates. Frames of reference, it seems to me, provide a crucial connection between an abstract mathematical Minkowski spacetime and the physical world which we wish to describe by means of it. Causality imposes some important constraints. If a spacetime line is to serve as the time-axis for a frame of reference, then that line must be time-like. It must constitute a possible trajectory for a material object. Frames of reference should – under some suitable idealization – be identifiable with possible laboratories in which experiments can be conducted and observations made. If there is a causal connection between events E_1 and E_2, or if they are causally connectable, then they cannot occur at the same time – they cannot be simultaneous and they cannot occur on a common spatial axis of any frame of reference.

It appears to be an historical fact that physicists and philosophers found little trouble in abandoning (with Einstein) the concept of a privileged rest frame defined in terms of its state of motion with respect to the luminiferous ether. Galilean relativity, along with "unsuccessful attempts to discover any motion of the earth relatively to the 'light medium'" (Einstein [1905], p. 37), seem to have convinced most that this concept is otiose. There is, accordingly, no argument about the element of convention in the designation of a particular frame as a rest frame. Physicists and philosophers have been far more reluctant to agree that one-way velocities are ill-defined quantities in the absence of a *definition* of simultaneity. It may be that rest and simultaneity are equally conventional concepts, from a physical standpoint, but that the conventional character of the one is much easier to swallow than is the conventional character of the other. If one wants to say that the adoption of a coordinate system is a matter of convention, and that the coordinates determine both what constitutes rest and what constitutes simultaneity in the frame of reference we are going to use, I do not see anything wrong. The significant and non-trivial character of these conventions arises with the claim that the frame of reference to which the coordinate system is attached is a physically admissible one. This entails that events lying on the time-axis must be causally connectable, and that events lying on a space-axis cannot be causally connected.

Suppose we choose some electrically neutral particle which is moving inertially (with no net forces acting upon it) through space. We choose the worldline of this particle as the time-axis of our frame of reference. The moving particle constitutes a causal process which has a definite path in spacetime. If the particle is moving through an electromagnetic field, and if disturbances in the field are impinging upon the immediate locale of the particle and emanating from its immediate locale, these disturbances constitute additional causal processes which determine a series of light cones all along the particle's trajectory. Suppose we pick a point on this trajectory and arbitrarily designate it $t = 0$. We now ask, what is the instantaneous state of the entire field (which may be fluctuating) at the time $t = 0$? According to the conventionalist, the only restriction is that the spacetime points of the field must not be causally connectable with one another, so any spacelike hyperplane will do. The hypersurface must be flat, and all of its points must lie outside of the light cone of the event $t = 0$. Nerlich denies this; he claims that a unique spatial hyperplane is determined by the choice of our t-axis. Both the conventionalist and the non-conventionalist agree, I take it, that the choice of the t-axis involves an arbitrary convention. The conventionalist maintains that this convention does not automatically determine the spatial axes as well; he holds that a further convention on simultaneity is required. The non-conventionalist says that the one convention determining the t-axis determines the spatial axes as well.

Winnie [1977a] has shown that there is an important sense in which the choice of a t-axis does determine the spatial axes as well. The behaviour of light signals suffices to define the relation of orthogonality. Thus, given the t-axis, we may demand that the x-, y-, and z-axes be mutually orthogonal, and that each of them be orthogonal to the t-axis. This uniquely determines the spatial axes apart from trivial rotations. The basic question would seem to come down to this: is the demand that the spatial axes be orthogonal to the time-axis a further convention, or is this simultaneity relation physically determined?

In September, 1977, a 'Symposium on Space and Time', guest-edited by John Winnie, appeared in the journal *Noûs*. It contained an article by David Malament, 'Causal Theories of Time and the Conventionality of Simultaneity', and my paper, 'The Philosophical Significance of the One-Way Speed of Light' [1977k]. This produced a startling juxtaposition. My article surveyed a number of widely diverse proposals for empirical ascertainment of the one-way speed of light, and I believe I was able successfully to show that each of them breaks down in some crucial respect. I concluded that it is not possible to ascertain the one-way speed of light by an experimental means which does not presuppose a definition of distant simultaneity or some tantamount convention. Malament demonstrated that the causal structure of Minkowski spacetime provides the basis for a unique explicit definition of simultaneity. He showed that, for any given timelike line which could serve as a time-axis in an inertial reference frame, there is one and only one equivalence relation (other than the vacuous relation or the universal relation) which is causally definable, and that the relation turns out to coincide with orthogonality. Certainly the relation of simultaneity (same time) must be an equivalence relation; moreover, in my view, causal relations are factual and non-conventional. Thus, Malament's result poses an extremely basic problem for the thesis of conventionality of simultaneity.

However, as Winnie noted in his editorial introduction ([1977b], p. 208), there is no direct contradiction between my thesis and Malament's result. Commenting upon the relationship between the two essays, he writes:

The situation is not presently clear for a number of reasons. First, the unique causal definability of standard simultaneity does not immediately imply that the adoption of an alternative (non-causally-definable) simultaneity relation must yield empirical predictions in conflict with those yielded by the adoption of standard simultaneity. Second, Malament's definability result relies essentially on global features of Minkowski spacetime, whereas some conventionalists (such as Reichenbach) have argued that no experimental facts, hence no *local* results, can yield a unique one-way velocity for a light path.

I must confess that I find the situation deeply perplexing; it raises fundamental problems concerning the relations between global theory and local experiments. Suppose — what is manifestly contrary to fact — that all of our physical observations and experiments supported the view that special relativity applies without restriction, so that by inductive extrapolation we would be justified in postulating Minkowski spacetime as the correct characterization of the spacetime structure of the world. Then, it seems to me right now (leaving open the option to revise this opinion in the light of further investigation) that we would be unwarranted in withholding factual status from simultaneity with respect to any given inertial reference frame. Malament's result may thus turn out to vindicate the fundamental thesis of Saunders and Norton that a unique simultaneity relation, which cannot be confirmed or disconfirmed in any direct experimental fashion, is nevertheless furnished by an overarching physical theory. Under the assumed conditions, of course, the theory would be empirically justified on the basis of observations and inductive extrapolations. Malament's result might also vindicate Nerlich's preference for L-frames over T-frames. It may be necessary, that is to say, to accept Malament's result as a refutation of the conventionality thesis with respect to unrestricted special relativity.

At time same time, it is essential to recall that this factual status of simultaneity is extremely fragile on account of its crucial dependency upon the global features of Minkowski spacetime. In a world – such as our own appears to be – where special relativity has highly restricted applicability, the factual status of simultaneity may evaporate. If this assessment of the situation is correct, I was simply mistaken in my claims about the conventionality of simultaneity in special relativity; nevertheless, the question of conventionality re-emerges in the context of more recent physical theory which restricts the applicability of special relativity. Malament's result deserves much serious study and philosophical reflection; the only point on which I feel at all confident is that I do not yet grasp its full import.

CAUSALITY AND SCIENTIFIC EXPLANATION

Glymour is right when, at the end of his essay, he characterizes me as a "Mechanical Philosopher", but I suspect he was wrong – perhaps it was wishful thinking! – when he called me the last. One piece of evidence is a postcard Bas van Fraassen sent to me shortly after he had read Glymour's paper. "If having none of modal realism, necessary connections, and occult powers requires [being classified as a Mechanical Philosopher]," he said, "I should like to join you in the 18th century!" Since the statistical probabilities must surely favor my demise before van Fraassen's, it seems likely that at least one of that possibly vanishing breed will survive me.

Joking aside, the characterization is apt because I reject the epistemic conception of scientific explanation, as sketched in "Comets, Pollen, and Dreams", and favor the ontic conception. I do not agree with Hempel's view that a scientific explanation of an event is an argument to the effect that the event-to-be-explained was to be expected by virtue of the explanatory facts. This view creates far too strong a bond between explanation and prediction. I do not deny that there is an important relationship between these two great aims of science; both, I believe, involve appeals to the laws of nature. There are, to be sure, some primitive forms of inductive inference which do not use laws, but scientific inference of any degree of sophistication relies heavily upon them. But here, on my view, the relationship ends. Prediction – or, better, reasoning to the unobserved – is an inferential activity; it involves the construction of logical arguments in an obvious and explicit way. Explanation is an entirely different sort of thing.

Scientific explanation is designed to provide understanding, and such understanding results, I believe, from *knowing how things work*. This is the sense in which I am a mechanical philosopher; understanding, I maintain, involves knowledge of the mechanisms. These mechanisms are, to a large extent, causal; hence, considerations of causality play a crucial role in the correct characterization of scientific explanations. Indeed, it seems to me, explanations in all scientific disciplines (with the possible exception of quantum mechanics) are causal in character. Leaving quantum mechanics aside for the moment, I accept Glymour's formulation of my view: "There is no such thing as statistical explanation *per se*, merely statistical evidence for (and, perhaps, statistical aspects of) causal explanations."

When Hempel published his first systematic treatment of statistical explanation [1962a], I felt immediately that somehow it was fundamentally on the wrong track. My original diagnosis was that the basic difficulty lay in Hempel's appeal to high probabilities when, in fact, statistical relevance is the key explanatory concept (Salmon [1965d], pp. 145–146). What I failed to appreciate at that time was that the *explanatory significance* of statistical relevance derives entirely from the fact that statistical relevance relations are indicative of underlying causal relations. I mistook the symptom for the real thing, and went on to try to develop a model of scientific explanation based upon the relation of statistical relevance. The result was the statistical-relevance (S-R) model of scientific explanation (Salmon, *et al.* [1971a], pp. 10–12). I recognized, of course, that causal considerations play an important role in scientific explanations, and that there is some sort of close relationship between statistical relevance and causality. I held out the hope that statistical relevance would turn out to be the basic concept, and that causality could be explicated entirely in terms of statistical relevance. This hope now seems ill-founded.

Hempel has never, as far as I can tell, appreciated the importance of the shift from high probability to statistical relevance in the theory of statistical explanation. Even in his 1976 Epilogue (Hempel [1977], Section 3.7), where he relinquishes the high-probability requirement for his own theory of statistical explanation, he sees the chief difference between the two approaches in the issue concerning the admissibility of explanations embodying low probabilities. Perhaps the main reason for the failure of adherents of the epistemic conception to appreciate the significance of statistical relevance lies in the fact that the epistemic conception does not accord an important place to causality. In the classic paper [1948] by Hempel and Oppenheim, it is assumed, more or less in passing, that deductive "covering law" explanations will be causal, but in his major subsequent monographic essay, Hempel explicitly denies that such explanations need be causal (Hempel [1965b], p. 352). One could, of course, modify the epistemic conception by requiring that the laws cited in deductive explanations be causal, but Hempel, at least, has not chosen to do so. This attitude is understandable. Causality is a feature which is crucial to the ontic conception, but there is little reason to give it any role in the epistemic conception. If, after all, the object of the exercise in formulating scientific explanations is to construct arguments which would have had predictive import if they had been available soon enough, then causal laws do not appear to have any advantage over non-causal laws. If a true law-statement can serve as a premise in a logically correct argument which has the explanandum statement as its conclusion, then no more is needed. It would be otiose to demand that such laws be causal. For a mechanical philosopher, in contrast, it is essential to appeal to causal relations, for they are constitutive of the mechanisms of the world.

What is often characterized as a mechanistic world-view is not very popular at present, largely because it is usually conceived in terms which are scientifically out-moded. A good – if somewhat primitive – example of a mechanistic world-picture is the atomism of Lucretius, Gassendi, or Laplace. In the latter part of the 19th century, mechanism was identified with the view of those physicists who tried to explain electromagnetic phenomena in terms of a mechanical ether. This viewpoint is epitomized in the famous remark of Lord Kelvin ([1884], p. 270): "I never satisfy myself until I can make a mechanical model of a thing. If I can make a mechanical model I can understand it." Thus, mechanism is often identified with the notion that explanations of physical phenomena are inadequate unless they are given in terms of levers, springs, pulleys, strings, wheels, gears, and deformable jelly. Construed in these terms, mechanism is scientifically anachronistic.

The difficulty with this late-19th-century version of the mechanical philosophy lies, not in the fundamental philosophical perspective, but rather, in its misconception of the *basic* mechanisms which are actually operative in the physical world. We realize today that the mechanical properties of such material objects as springs and gears require explanation in terms of the fine structure of macroscopic objects. We recognize, for example, that electrostatic forces play a crucial determining part in explaining such structures. Because of Maxwell's electromagnetic theory and Einstein's special theory of relativity, we now believe that the electromagnetic field has a fundamental physical reality. Electromagnetic phenomena are not to be explained in terms of the behavior of a mechanical medium. The situation is just the reverse. The mechanical properties of the macroscopic objects with which Lord Kelvin made his models are to be explained in terms of the properties of fields and of the waves which propagate through such fields. As I tried to say, somewhat sketchily, in my discussion of causal processes in 'Comets, Pollen, and Dreams', the propagation of such waves through these fields is a fundamental causal mechanism in our universe. We have to change our mechanistic view from the crude atomism which recognizes only the motions of material particles in the void to a conception which admits such non-material entities as fields, but for all of that, it is still a mechanistic world-view. Old-fashioned materialism is now untenable, and so – in all likelihood – is determinism, but the mechanical philosophy lives on. However, if it is to survive it will have to take into account the great mechanical theory of the 20th century – quantum mechanics – as the essays by Forge, Glymour, and van Fraassen all point out. I shall return to this topic below.

The idea that causality plays an important role in scientific explanation is by no means new; in many non-philosophical contexts, I believe, the phrases "finding the cause of an occurrence" and "finding the explanation of an occurrence" are taken as nearly interchangeable. Unfortunately, it seems to me, those philosophers who have most strongly insisted upon the causal aspects of explanation have not made sustained efforts to provide adequate explications of the required causal concepts. The ontic – or mechanical – approach to scientific explanation cannot succeed without such explications.

Traditional treatments of causality have usually proceeded as if there is one basic causal concept, namely a relation of some sort which obtains between two events when one is the cause of the other. I am inclined to believe that there are two fundamental causal concepts – causal *propagation* and causal *production*. Causal propagation takes place within causal processes.

The fact that a causal process is *capable* of transmitting a mark indicates that such a process is *actually* transmitting or propagating its own structure. By virtue of this latter fact, we may say that causal processes transmit structure, order, information, energy, and causal influence. As I tried to argue in detail in [1977f], it seems to me that the sort of causal propagation which takes place within causal processes *is* the connection between cause and effect which was sought by Hume and many other philosophers.

In adopting this approach to causal connections — which are *not* the kinds of "necessary connections" which Hume proscribed — I am quite deliberately taking processes rather than events as the basic entities treated by causal theory. This approach is scientifically justified, I believe, within the context of the special and general theories of relativity, by the fact that the space-time structures treated by these theories can be defined entirely on the basis of causal processes. For a variety reasons, some of which I have sketched in (Salmon [1980a]), it seems to me that all attempts to work out theories of causality in terms of such relations as sufficient condition, necessary condition, and statistical relevance among discrete events are inadequate. They fall short precisely because they fail to take account of the physical connections provided by causal processes.

In view of this consideration, I suggest that, instead of looking at the world initially as a manifold in which many *events* are located, we look at the world primarily as a manifold in which *processes* transpire. The most obvious difference between events and processes is that events are spatio-temporally localized, while processes are extended in time. At first glance, we have no distinction between causal processes and pseudo-processes, but we can note that many processes intersect with one another. If two processes which intersect both suffer modifications which persist beyond the point of intersection, and which are correlated with one another in specific ways, then we can say that a *causal interaction* has occurred, and that the two intersecting processes are both causal processes. Such an interaction is an event — indeed, it is a marking event. In this marking event, each process is marked by virtue of its interaction with the other. In the interaction, each process *produces* a modification in the other, and the modifications are *propagated* by the processes. This mutual modification is a sufficient condition for the two processes both being causal, but it is not necessary. Two light rays, for example, both of which are bona fide causal processes, can intersect without interacting. This is a characteristic feature of waves.

The kind of correlation which obtains between the modifications in two interacting processes is characterized in terms of a statistical structure which

I have called an *interactive fork*. My attention was first directed toward interactive forks by van Fraassen in the discussion to which he refers in his essay in this volume. In [1975d], I attempted to bring out explicitly the role of causal processes and causal forks in the theory of scientific explanation. I laid particular stress upon Reichenbach's *principle of the common cause*, and upon the statistical structure which he called a *conjunctive fork*.

In his [1977], van Fraassen sharply challenged the claim that conjunctive forks correctly characterize all common cause situations, and in the above-mentioned subsequent discussion he forced me to think much more carefully about forks. I am extremely grateful to him on two counts. First, he has taken the common cause principle seriously. It is a fundamental explanatory principle, but it has been ignored by virtually all philosophers who deal with scientific explanation. Second, he has made me realize – what should have been obvious – that there are at least two kinds of forks, *interactive* and *conjunctive*. Reichenbach's formulation of the common cause principle, which recognizes only conjunctive forks, is clearly untenable. I had adopted this version until van Fraassen showed its inadequacy.

Reichenbach's attention was drawn to conjunctive forks ([1956], Section 19) because they appear to exhibit a temporal asymmetry. The event at the vertex of the fork is always a common cause, but never a common effect. Interactive forks, I believe, also play a fundamental role; indeed, I suspect that they characterize the fundamental interactions in nature, but they lack temporal asymmetry. Interactive forks are involved in the production of changes of order or structure. Conjunctive forks hold the key to the question of why events must be explained in terms of preceding conditions rather than subsequent ones. Forks of both types are indispensable for any adequate formulation of the principle of the common cause. I have offered somewhat fuller discussion of the issues surrounding forks of these two types in [1978c].

There is another debt of gratitude which I owe partly to van Fraassen. One of his students, Ellis Crasnow, provided an example which shows that the statistical relations embodied in Reichenbach's definition of the conjunctive fork are inadequate, in some cases, to discriminate between specious common causes and genuine ones. The difficulty can be overcome, I think, by making appropriate stipulations about the causal connections among the events which constitute the fork in terms of the causal processes which provide these connections. This is one important reason, among several, for believing that causal relations cannot be defined in terms of relations among discrete events without making reference to connecting processes. I have discussed some of these (including Crasnow's example) at some length in [1980a].

To anyone who has ever thought about the matter, it should be obvious that the principle of the common cause cannot be an a priori truth. If there was any doubt, van Fraassen's clever examples surely settle the question. It seems, rather, to represent an extremely important and fundamental fact about the physical world in which we happen to live. Indeed, Reichenbach argues — correctly I think — that his version of the principle has essentially the same status as the second law of thermodynamics, namely, that highly ordered states of isolated systems do not arise spontaneously by chance. It is not difficult to imagine a world in which neither the common cause principle nor the second law of thermodynamics holds true. If, as I maintain, the common cause principle is a basic explanatory principle, that tells us something important about the theory of scientific explanation.

The classic Hempel-Oppenheim article [1948] dealt, according to the title, with the *logic* of scientific explanation, where the term "logic" must be taken to include both syntactic and semantic features. If truth, deductive validity, and lawlikeness are semantic concepts, then the account of deductive-nomological explanation offered in that paper was semantical. It is my seriously considered opinion that little, if anything, can actually be accomplished within the realm of semantics alone to provide an adequate characterization of scientific explanation, for what constitutes a satisfactory explanation will have to depend upon certain general features of the world in which we live. I am convinced that the effort to find an a priori "logic of explanation" has had a deleterious effect upon discussions of scientific explanation in the last three decades, and I consider it a defect of the epistemic conception that its characterization of explanation can be given in semantic terms.

The ontic conception of scientific explanation is patently non-semantical, for its characterization of explanation involves aspects of the world which are manifestly not a priori. To assert that the world has mechanisms, and to make any claims about the nature of these mechanisms, obviously involve factual propositions. To maintain that there are spatio-temporally continuous causal processes, that they interact with one another, and that there are two types of causal forks involves further factual assertions. To claim that causal processes and causal forks have explanatory significance is, again, to make factual statements about the world. Two features of the epistemic conception of scientific explanation, which I consider serious defects, can be traced directly to this source. First, the epistemic conception has no non-arbitrary way of incorporating the requirement that events be explained on the basis of preceding conditions rather than subsequent ones. Second, the epistemic conception does not reflect any difference in explanatory power between causal and non-causal laws, and it has no role for causal mechanisms (see Salmon [1975d], [1977g]). Hempel's discussion of these two issues in ([1965], pp. 347–354) reveals these problems quite clearly. The a posteriori status of these considerations follows from the factual nature of the temporal and causal structure of the world.

Someone much cleverer than I might be able to formulate the ontic conception in a way which would make it applicable to all possible worlds, but I feel quite sure I cannot. I do not know, for example, what would qualify as a "mechanism" or a "causal connection" in a world radically different from our own. To formulate an explication of scientific explanation which would have *a priori* applicability to all possible worlds would require such a feat.

The same moral also applies, of course, to theories of scientific explanation intended for applicability to this world — they cannot be a priori. This point has important bearing upon the issues discussed in this volume. In 'Comets, Pollen, and Dreams', I argued that we must provide an account which allows room for functional explanations, because they do appear to figure importantly in various branches of science. The same consideration applies to the closely related point about explanations of low-probability events. It would be a grievous error to adopt a Procrustean theory which rules out explanations of these types on some a priori basis. It strikes me, incidentally, that the modal conception, in categorically precluding statistical explanations of particular occurrences, suffers greatly from this defect.

Perhaps the best way to put the matter is to insist that philosophical theories of scientific explanation should be constructed piecemeal, partly in terms of accepted explanatory practice in various branches of science, and partly in terms of our own general conceptions of the fundamental and pervasive features of the world. I do not mean to suggest that our theories must accommodate every explanation which is advanced by any scientist, or every explanation which is accepted by a large number of scientists. I mean, rather, that certain types of explanations, such as statistical explanations or functional explanations, should be taken seriously — and subjected to careful philosophical analysis — if explanations of these types are advanced and accepted in large and significant regions of scientific practice. Although (I am optimistic enough to believe) we have made considerable progress in characterizing scientific explanation outside of the quantum domain, it remains to be seen what sorts of explanations are admissible in quantum mechanics. Even if quantum mechanics should require patterns of explanation radically different from those I have characterized, that fact would not undermine their legitimacy for non-quantum-mechanical contexts. Since we are not attempting to formulate a "logic of scientific explanation", there is no presumption that any particular pattern or model can attain universal applicability.

The a posteriori character of the common cause principle is effectively exhibited by van Fraassen's fanciful macroscopic examples. No logical inconsistencies are involved in their descriptions. Reichenbach recognized the same fact when he remarked that there is no *logical* obstacle to conjunctive forks which have common effects at their vertices instead of common causes. In attempting to understand the case of Simpson's paradox given by van Fraassen in Section 5, I tried to see whether his probability relations could be realized in a non-fanciful example. We require:

$$P(x) = P(\overline{x}) = \tfrac{1}{2}, \qquad P(A) = P(B) = \tfrac{1}{3}, \qquad P(AB) = \tfrac{1}{6}$$
$$P(A|x) = P(B|x) = \tfrac{2}{3}, \qquad P(AB|x) = \tfrac{1}{3}$$
$$P(A|\overline{x}) = P(B|\overline{x}) = P(AB|\overline{x}) = 0.$$

It was not too difficult to contrive a chance set-up which, by deliberate design, would yield these values. Suppose that we have cubes and tetrahedrons made of gold and silver. Let A stand for the event of finding a gold cube, and B the event of finding a gold tetrahedron. We have two cabinets, each of which has three drawers. In one of the cabinets, the first drawer contains a gold cube and a silver tetrahedron ($A\overline{B}$), the second drawer contains a silver cube and a gold tetrahedron ($\overline{A}B$), and the third drawer contains a gold cube and a gold tetrahedron (AB). In the other cabinet, each of the drawers contains a silver cube and a silver tetrahedron ($\overline{A}\overline{B}$). The player tosses a fair die. If side 1, 2, or 3 shows (event x), he opens the first, second, or third drawer, respectively, of the first cabinet. If side 4, 5, 6 shows (event $\overline{x}$), he opnes the first, second, or third drawer, respectively, of the second cabinet. Evidently, the foregoing probability relations are satisfied. Since the toss of the die precedes the opening of a drawer, we may speak of the result of the toss as the "cause" of finding a gold or silver cube or tetrahedron.

Now, this may represent merely a failure of my imagination, but I was unable come up with a situation, which had not been contrived especially for the purpose, which would exhibit these values if x stands for a cause. If, however, x stands for an effect, it seems fairly easy to find examples which are not too far-fetched. Imagine a college in which either passing a mathematics course (A) or passing a logic course (B) is a requirement for graduation. Assume that all students who do not graduate fail to fulfill this math/logic requirement. Some students, of course, fail to graduate for other reasons

— e.g., because they cannot pass English composition. Students of this sort never even get to the mathematics or logic courses. Assuming an attrition rate of 50%, assuming that half of the students who pass a mathematics course also sign up for a logic course which they pass, and assuming that half of the students who pass a logic course also enroll in a mathematics course which they pass, we have all of the probability relations stipulated by van Fraassen.

Consider another example. Suppose we have a theatrical troupe in which the leading lady is sometimes ill (A) and the leading man is sometimes ill (B). If either of them is ill, the evening's performance is cancelled (event x). We need not assume that half of the performances are cancelled; indeed, if we set $P(x) = \frac{1}{10}$, the point of van Fraassen's examples is magnified. It is important to remember that A and B cannot be independent; they are positively correlated. This feature of the example is realistic; since many illnesses are contagious, we should expect the illness of one of the players to increase the probability of the illness of the other.

Over the past several years, I have given a good deal of thought to the relationship between causal (or temporal) asymmetries and various statistical structures such as conjunctive forks. What I have learned, primarily, is that these questions are extremely complicated and subtle. At some time in the near future, I hope to be able to present a reasonably comprehensive treatment of them, but at the moment, I cannot. Thus, I must confess, I do not have any adequate response to van Fraassen's ingenious examples, but I know that consideration of such cases can be highly instructive.

The principle of the common cause operates sometimes as an explanatory principle and sometimes as a principle of inference. Van Fraassen's essay brings out the importance of distinguishing these two uses. Since the statement that persistent coincidences occur only in the presence of common causes does not enjoy *a priori* status, it does not seem reasonable to adopt, as a *primary* inductive method, a rule which directs us to infer the existence of a common cause in such circumstances. If we are to avoid begging the question, we must, it seems to me, justify our primary inductive rule without appeal to such factual assertions as the principle of uniformity of nature or the principle of the common cause. As I argued in the section on induction and probability, it seems possible to provide a *vindication* of the rule of induction by enumeration without invoking any factual claims. If this justification is satisfactory, then we can use our primary inductive rule to *validate* other secondary rules of inference. The rule — *whenever certain kinds of coincidences occur persistently, infer that a common cause is present*

– has certainly achieved a great deal of success in science and in everyday life. This qualifies it, in van Fraassen's terms, as an excellent tactical maxim, at least until we get into the quantum domain. Whether it will eventually prove applicable in microphysics is problematic. I certainly agree with van Fraassen's view that adoption of that principle cannot be considered a necessary condition of rationality, even though it clearly is rational to adopt it in some domains of inquiry.

Regardless of its status as a principle of inference, the principle of the common cause *may be* a fundamental principle of explanation *in our world*. I am not sure whether it has this status in quantum mechanical contexts. It seems to me that the Einstein-Podolsky-Rosen problem nicely illustrates the difficulties surrounding explanation in the context of quantum mechanics (see d'Espagnat [1979]). This problem concerns certain kinds of correlations between physical systems which have undergone some sort of physical interaction and subsequently been separated. Although there are many aspects of this problem which I do not pretend to understand, one aspect seems clear. Any satisfactory explanation of the observed correlations is bound to accord an essential role to the earlier physical interaction. In (Salmon [1978c]), I discussed Compton scattering as an example of an interactive fork. One of the features of that interaction is that energy is conserved. In certain versions of the Einstein-Podolsky-Rosen problem, the crucial feature of this interaction is conservation of angular momentum – for example, the total angular momentum of the correlated systems is zero. *How* nature manages to arrange for the conservation of angular momentum is something of a mystery, given what we believe about quantum phenomena, but a causal interaction governed by a conservation law is not, in itself, mysterious. Causal mysteries surround the Einstein-Podolsky-Rosen problem, but violation of the principle of the common cause is not *obviously* among them.

Suppose, however, that quantum theory implies genuine violations of the principle of the common cause. We would *not* be justified, in my opinion, in stating flatly that the quantum theory must be wrong, and that it must be replaced by a hidden variable theory which allows us to infer the existence of common causes. If the common cause principle is seen as a demand for hidden variables under these circumstances, then I think it should be rejected. As a principle of *inference*, it cannot legitimately be forced upon the quantum domain. At the same time, two stances are possible with respect to *explanation*. First, we may try to find other acceptable patterns of explanation which do not embody the common cause principle. Second, we may say that phenomena which involve violation of the common cause principle are real enough, but they may simply be incomprehensible to us. As van Fraassen remarks, "It was therefore an unprecedented step for the Copenhagen school to take the position that *yes*, there were positive statistical correlations in nature that quantum theory could not eliminate by exhibiting preceding conditions relative to which those correlations disappear, but *no*, the theory was not incomplete." It is possible to admit those facts about the quantum theory and its relation to nature, and still to deny that the quantum theory can explain such phenomena. Perhaps a major source of the uneasiness which many theorists felt with respect to the Copenhagen interpretation is that it seems to provide enormous predictive and inferential power, but that it is altogether lacking in explanatory power.

A recent *Scientific American* article, 'The Quantum Theory and Reality' by Bernard d'Espagnat [1979], brings out this issue with admirable clarity. At the outset, d'Espagnat focuses explicit attention upon the problem of explanation:

Any successful theory in the physical sciences is expected to make accurate predictions. . . . From this point of view quantum mechanics must be judged highly successful. As the fundamental modern theory of atoms, of molecules, of elementary particles, of electromagnetic radiation, and of the solid state it supplies methods for calculating the results of experiments in all these realms.

Apart from experimental confirmation, however, something more is generally demanded of a theory. It is expected not only to determine the results of an experiment but also to provide some understanding of the physical events that are presumed to underlie the observed results. In other words, the theory should not only give the position of a pointer on a dial but also explain why the pointer takes up that position. When one seeks information of this kind in the quantum theory, certain conceptual difficulties arise [p. 158].

These difficulties are directly related to the common cause principle:

The need to explain observed correlations is so strong that a common cause is sometimes postulated even when there is no evidence for it beyond the correlation itself. Whether or not this procedure can always be justified is a central issue [p. 160].

d'Espagnat does not have a real solution to this perplexity, but he does categorically reject one alternative:

One conceivable response to the distant-correlation experiments is that their outcome is inconsequential . . . the results are merely what was expected. They show that the theory is in agreement with experiment and so provide no new information. Such a reaction would be highly superficial [p. 181].

In this judgment, I believe (as Forge presumes I would [p. 227]), d'Espagnat is altogether correct. It is worth noting that the alternative he rejects is precisely the explanation which would be supported by the epistemic conception. If the theory correctly predicts the experimental outcomes — statistically if not deterministically — then the theory *ipso facto* explains them. At this juncture — rightly or wrongly — I experience intense dissatisfaction with the epistemic conception's answer. I am not quite sure what posture to adopt, and I do not believe a hasty decision is in order. We need to know a great deal more than we do now about the nature of quantum mechanical explanation before we can make a reasonable choice between these alternatives.

In [1978c], I sketched a sort of common cause argument in support of scientific realism, using two ways of ascertaining Avogadro's number as an example. This example is of first importance, both historically and philosophically. As one elementary text in quantum mechanics puts it, "Avogadro's number N_0 is the link which connects microphysics with macrophysics" (Wichmann [1967], p. 20). If this statement is (as I believe) correct, then it makes good sense to look at this example in connection with the philosophical thesis of scientific realism. The point is reinforced historically by the fact that physical scientists who doubted the existence of molecules and atoms around the turn of the century were (with the outstanding exception of Ernst Mach) convinced of their reality by the investigations of Avogadro's number by Jean Perrin at the beginning of the twentieth century. It seems to me quite remarkable that — to the best of my knowledge — this crucial example played no role in philosophers' discussions of this issue prior to Gardner [1979], who does not consider what is, in my opinion, *the* key aspect of the argument. If we want to know what reasons can be given for asserting the existence of unobservable entities, the least we can do is to examine the evidence scientists found compelling. It is almost equally remarkable that Clark Glymour, who first called my attention to the excellent account of Perrin's work in Mary Jo Nye's *Molecular Reality*, should dismiss my appeal to this argument as unfounded numerology!

Perrin's work constitutes a landmark in the history of science, and one which carries philosophical implications of great moment. It is, I believe, *the first time* that scientists found compelling evidence for the existence of micro-entities. There had, of course, been philosophical and scientific speculation about the existence of atoms for more than two millenia, but prior to Perrin's work, scientists could reasonably remain skeptical about these entities. Wilhelm Ostwald, a vehement opponent of the atomic or molecular hypothesis as late as 1906, declared in the preface of the fourth edition of his treatise on physical chemistry in 1908:

I have satisfied myself that we arrived a short time ago at the possession of experimental proof for the discrete or particulate nature of matter – proof which the atomic hypothesis has vainly sought for a hundred years, even a thousand years (Quoted by Nye [1972]:51).

What was the nature of the evidence which brought forth such strong scientific conviction even from skeptics? It was *not* just one ingeniously designed and masterfully conducted experiment which provided an accurate determination of Avogadro's number; it was, instead, a large number of different methods for ascertaining the same number, and the fact that they all agreed with one another within the limits of experimental error. In his popular book *Atoms*, Perrin provides a table containing 13 distinct ways of ascertaining Avogadro's number, involving such diverse phenomena as gaseous viscosity, Brownian movement, the blueness of the sky, black body spectra, and radioactivity. His list does not include the electro-chemical method I discussed ([1978c], p. 698), which constitutes still another independent approach. Concerning the data in this table, Perrin writes:

One is seized with admiration before the miracle of such precise agreements coming from phenomena so different . . . this gives a probability to molecular reality really bordering on certitude. . . . The atomic theory has triumphed. Numerous just a short time ago, its adversaries – finally won over – renounce one after another the suspicions which were legitimate and no doubt useful for a long time (pp. 206–207).

If the observable characteristics of matter exhausted its real properties, then the diverse experiments leading to values for Avogadro's number would truly be independent. To one who knows nothing about the fine structure of matter, these determinations appear to be independent. Once the existence of such unobservable entities as atoms and molecules is acknowledged, they are, of course, no longer independent; they are merely distinct ways of investigating the same kinds of entities. What would be a miraculous coincidence if atoms and molecules are unreal becomes an explainable agreement given the reality of these unobservable entities. If one were to argue from these data that every mole of gas contains 6.022×10^{23} angels dancing to celestial music, that would be numerology, as Glymour charges. However, atoms and molecules, in contrast to angels, have structural and causal properties which make them useful in explaining all manner of physical phenomena. Perrin's argument for molecular reality is strong precisely because it appeals to underlying causal processes and because it has the form of a

common cause argument. The above-mentioned books by Perrin and Nye
should constitute required reading for all philosophers who want to discuss
scientific realism. This is one case, I think, in which attention to the history
of science pays rich dividends for the philosopher of science. Among other
things, it helps one appreciate concretely what was, I believe, the force of
J. J. C. Smart's famous argument from "cosmic coincidence" ([1963], p.
39).

Glymour concludes his discussion by saying that many of his points are
"little more than cavils or qualifications or suggestions for further work."
I am glad to take them in that spirit. It seems to me that the points he makes
about factor analysis and other causal considerations in the behavioral sciences
should fit well with the kind of theory of explanation which I am attempting
to develop. If revisions or modifications of detail are required, they should
be duly incorporated. I see little reason to despair about the ability of my
approach to handle in appropriate causal fashion such items as "Dalton's
atoms, Newton's gravitational force, Neptune, Vulcan, black holes, neutrinos,
etc.". I am grateful for his many examples, and for his hints about how
they bear upon a theory of scientific explanation.

One criticism, however, strikes deep in Glymour's opinion – the issue
about action-at-a-distance. Here, I think it is crucial to divide the problem
into two parts, namely, the quantum domain and other scientific realms.
Glymour has two reasons for objection to the rejection of action-at-a-distance
outside of the quantum domain. One has to do with Newtonian gravitation;
the other is the argument about big things being made out of little things.
Let me consider both.

First, although it may have been reasonable to admit action-at-a-distance
in the early post-Newtonian era, I think one can make a good case for rejecting it in macrophysics today. The work of Faraday, Maxwell, and Einstein
has given us good grounds to regard electromagnetic fields as real, and to
regard the propagation of deformations in the field as bona fide causal
processes. Current cosmological theory seems to accord the same status to
gravitational fields, and to the propagation of deformations in these fields.
While it may be that no one has, as yet, succeeded in detecting gravitational
waves directly, gravitational radiation does play an important role in the
explanation of astrophysical phenomena. Action-at-a-distance has no place,
I believe, in the world picture which results cumulatively from Newton,
Maxwell, and Einstein.

Second, although big things are indeed made out of little things, we must
be careful in making inferences from microphysics to macrophysics. It may

be that certain macro-phenomena – e.g., superfluidity – are manifestations of quantum mechanical properties of matter. It may be that action-at-a-distance occurs at the quantum mechanical level. Nevertheless, I do not know of any macrophysical phenomenon which can reasonably be interpreted as macroscopic action-at-a-distance arising from quantum mechanical action-at-a-distance operating in the microscopic domain.

It seems that quantum mechanical discontinuities are confined to microscopic phenomena. If, in a two-slit electron diffraction thought experiment, we treat the electron as a wave, the only causal anomaly – the only sort of action-at-a-distance – occurs when the electron interacts with a localized detector. But when we look at the pattern which is built up out of many of these localized electron interactions, the overall result looks just like the wave we used in the first place. The effects of action-at-a-distance average out when large aggregates are involved. It is by no means evident that quantum discontinuities pose any threat to the *macroscopic* principle of action-by-contact.

Although a great deal remains to be done to fill in details, and perhaps to make revisions in the light of specific criticisms, I do not find grounds for serious dissatisfaction with the general features of my statisti-causal account of scientific explanation as it applies outside of the quantum domain. I am, however, painfully aware of the difficulties involved in trying to develop any adequate theory of quantum mechanical explanation. In view of remarks made above, it should be obvious that I would be untrue to my own basic philosophical convictions if I were to try to impose, in some a priori way, either the common cause principle or the requirement of action-by-contact as conditions of adequacy upon quantum mechanical explanations. The criticisms offered by Forge, van Fraassen, and Glymour show that there are genuine unresolved problems associated with explanation in microphysics.

Up to this point, I have said very little about John Forge's excellent essay, mainly because I agree whole-heartedly with almost everything he says. His assessment of philosophical accounts of explanation in terms of their applicability to serious scientific examples is altogether laudable, and his discussion of the consequences of the theory of errors is original and valuable. If I correctly understand him, we both adhere to an ontic conception and we attribute similar importance to causal relations. The major problem concerns the way in which causality is to be extended into the quantum domain. This is a deep and difficult subject.

Near the end of his essay, Forge offers a suggestion which is appealing in its simplicity: "A causal process is one which is governed by scientific laws (theories), whereas a pseudo-process is not." Unfortunately, it seems to me, this solution is not satisfactory, for it fails to do justice to cases of the sort exemplified by the Einstein-Podolsky-Rosen problem. For a mathematically precise and theoretically sound treatment of one such experiment, see Cantrell and Scully [1978]. Although the laws of quantum mechanics govern the behaviors of the correlated systems, it is hard to see how such laws, *ipso facto*, introduce causal components. The conceptual problems raised by d'Espagnat [1979], discussed above, seem to me to constitute severe obstacles to Forge's positive proposal.

I *do* agree with Forge in the claim that, if causal considerations are pertinent in quantum mechanical contexts, we may need to revise the very concept of causality in some appropriate – perhaps drastic – way. I am not sure, however, that the notion of the mark must be abandoned altogether. A microphysical process, such as a moving particle, can be "marked" in the sense that one of its physical parameters can be altered. In detecting the positron – as Dirac ([1978], p. 18) mentioned in a 1975 lecture in Sydney – Anderson placed a lead plate in its path to reduce its kinetic energy. By observing the shape of the path, Anderson was able to show that the particle producing the track in his cloud chamber was a positron rather than a negative electron. The basic idea of marking comes from the concept of the interactive fork. If two processes intersect, and both are modified in ways which persist beyond the point of intersection, it seems permissible to say that both were marked in the interaction. Such interactions are common within quantum mechanics. Moreover, as is shown by the case at hand, microparticles do have trajectories. Although the trajectory of an electron or positron in a cloud chamber is not a proper classical trajectory which associates a precise spatial position with each precise temporal moment, allowing suitable uncertainities in position and velocity does result in a smeared-out trajectory. Moreover, it would be hard to deny that such processes transmit causal influence. The picture tube in any television receiver provides convincing evidence.

A great deal has been said, both in the philosophical and in the scientific literature, about the difficulties of extending causality into microphysics. One version of Bohr's principle of complementarity states that the space-time description and the causal description of a given microsystem are incompatible. Reichenbach [1944] formulated a principle which asserts that causal anomalies cannot be totally eliminated from quantum mechanics (see also Jones [1977]). Most such discussions are based upon the Heisenberg-Schrödinger-Born non-relativistic quantum theory, which is by no means the deepest microphysical theory available today. Before coming to any firm conclusions about causality, we should give careful consideration to quantum electrodynamics, the theory which covers interactions between charged particles and photons. This is, in some respects, the most satisfactory physical theory yet to be devised. R. P. Feynman, in a systematic series of articles [1949a, 1949b, 1950] – further elaborated in Feynman [1961a, 1961b] – attempted to provide a space-time formulation of quantum electrodynamics.

This theory makes obvious departures from causality as conceived in the context of classical physics, e.g., it admits pair production and pair annihilation of particles; but since mass is not conserved in relativity theory, such occurrences do not even count as causal anomalies in relativistic quantum theory. Because energy is not strictly conserved in quantum mechanics, we have to contend with flocks of virtual particles, but their scope is severely limited by the uncertainty relation $\Delta E \Delta t \geqslant \hbar/2$. In Feynman's formulation, electrons *can* be said to scatter backward in time from regions of electromagnetic interaction, but with pair production and annihilation, it is *not necessary* to adopt this interpretation. An overall causal picture seems to emerge. Causal interactions occur in small regions of spacetime. If causal anomalies are involved in these interactions, they are localized by the Heisenberg uncertainty relation. None of these interactions involves action-at-a-distance; indeed, in quantum field theory, a condition of Lorentz-invariance, which precludes causal interactions between regions which sustain a space-like separation from one another, is imposed. Outside of localized regions of interaction, particles (including photons) behave as fairly reasonable causal processes. This picture of microphysical processes and interactions is not altogether at odds with the macrophysical account, sketched above, of causal processes and interactions. A detailed up-to-date treatment of the Feynman approach – which applies to the strong, weak, and gravitational forces, as well as the electromagnetic force – can be found in Scadron [1979]. The basic features of microphysical causality are treated, sometimes explicitly and sometimes implicitly, at considerable depth.

Quantum electrodynamics is not without difficulties, but it is a far deeper theory of microphysical mechanisms than was the early quantum mechanics of the mid-1920's. More recently, quantum chromodynamics has come to the fore as a theory of the interactions of hadrons. It bears some resemblances to quantum electrodynamics, but it is not nearly as fully developed. On the basis of my severely limited understanding of both theories, I fail to see any indication that quantum chromodynamics will not fit the general picture of causal processes and interactions I sketched in connection with quantum electrodynamics. Thus, it does not seem unreasonable to think – in the *most* tentative way – of microphysical explanation in terms of causal processes and causal interactions. Causal explanation may turn out, after all, not to be "a relic of a bygone age".

Whether such principles as spatio-temporal continuity of causal processes are ultimately tenable depends upon the future of physical theory, a topic on which I am *not* qualified to speculate. The situation regarding action-at-a-distance is not altogether clear. Perhaps the very idea of a four-dimensional spacetime is legitimate only for the macrocosm. Steven Weinberg, who received the Nobel Prize in 1979 for his work on unified field theories – and who *is* entitled, if anyone is, to speculate on such questions – recently offered some pertinent considerations (Weinberg [1979]). Speaking of a future theory which would embrace the strong, weak, electromagnetic, and gravitational forces, he remarked, "I think it is reasonable to suppose that . . . the ultimate physics which describes nature on smaller scales will in fact be of a geometric nature." It may turn out, however, that the space described by this geometry will not be of the familiar variety. " . . . I strongly suspect," he continues, "that ultimately we will find that the four-dimensional nature of spacetime is another one of the illusory concepts that have their origin in the nature of human evolution, but that must be relinquished as our knowledge increases" [p. 46].

We do, as Glymour says, live in mad times – and exciting times. Perhaps the reason no one has an adequate philosophical theory of explanation for quantum mechanics is that we do not yet know enough about the spatio-temporal and causal structure of the microcosm. In any case, to develop a philosophical understanding of quantum mechanical explanation is a premier challenge to contemporary philosophy of science.

A PLEA FOR PATIENCE

Fundamental philosophical problems are not usually solved quickly, easily, or with a single stroke. Zeno's paradoxes of plurality and motion furnish a good example. From antiquity to the modern period, they posed deep problems for our understanding of space, time, and motion. They were, to a very limited extent, resolved by the invention of the infinitesimal calculus; but, to a far greater extent, they posed serious foundational problems with respect to the calculus itself. By the end of the 19th century, when the foundations of the calculus had been put into proper order – approximately 2400 years after these paradoxes were first propounded – the means to their adequate resolutions were *fairly* well in hand. Not before the middle of the 20th century, however, did we have anything like a *complete* resolution of these ancient riddles. Whether we now have fully satisfactory treatments of them may be a topic of serious dispute, but either way, there can be no doubt that our understanding of space, time, and motion has been almost immeasurably deepened by those who have struggled with Zeno's paradoxes. And, it should be added, such understanding could not have emerged without important advances in mathematics. Many pieces had to be put together. It took a long time to do so, but it was well worth the trouble (Salmon [1975a] or [1980b], Chap. II).

Another closely related example from antiquity immediately suggests itself. Certain problems concerning geometrical knowledge were set forth by Plato, and their consideration played a major role in the development of his philosophy. "Let no one unversed in geometry enter my doors," said the sign over the porch of his Academy according to legend. Centuries later, the non-Euclidean geometries were discovered, precipitating one of the greatest intellectual revolutions of all time. The attention which these developments drew to the foundations of geometry, and to the problem of the applicability of geometrical systems to the physical world, resulted in a far clearer understanding of the nature of geometrical knowledge and of the notion of physical space than we ever could previously have attained. This achievement has had profound consequences for epistemology in general, for geometrical knowledge played a pivotal role in the problem of the synthetic *a priori*. Again, the philosophical road was long and hard, and the contributions of mathematicians and physical scientists were necessary prerequisites to philosophical progress. In this instance, in contrast to Zeno's paradoxes, we can be sure that the task is far from finished, and that we have much to learn about the nature of space. It does not follow that nothing has been achieved.

Compared with the foregoing examples, the problems relating to causality and induction which we have inherited from Hume are recent arrivals upon the philosophical scene. They, too, are deep and difficult problems, and the fact that we do not yet have fully adequate answers should occasion no great surprise. Nor should it lead us to view them impatiently as pseudo-problems. It is my personal view — which, I realize, is probably a minority opinion — that we *have* made some significant progress in dealing with them; indeed, it seems to me, the contributions to this volume bear witness to such progress. If this assessment is accurate, we have just cause for pleasure and ample reason to persist in our efforts. The reward will be a deeper comprehension of scientific knowledge and scientific understanding, and a vindication of the conviction that science can provide rational predictions and bona fide explanations of natural phenomena. In times of rampant irrationalism the ideals of the Enlightenment provide the only viable antidote.

NOTE OF THANKS

It is quite impossible for me to express adequately my gratitude to those who have been responsible for this volume. In the first instance, I owe warm thanks to the University of Melbourne and its Department of History and Philosophy of Science for the opportunity to spend three stimulating months in Australia, which enabled me to engage in fruitful interactions with faculty and students at Melbourne and at a number of other Australian universities. I am especially grateful to John Clendinnen and Rod Home – old friends from my days in History and Philosophy of Science at Indiana University – who were instrumental in making the practical arrangements for the visit. I should like to express appreciation to Ormond College (Melbourne) and to its former Master, Davis McCaughey, and Vice-Master, Rachel Faggetter, for providing extremely congenial living arrangements during the stay in Australia. The lives of my wife Merrilee and me were vastly enriched by learning the true meaning of Australian hospitality – both personal and academic.

The Australasian Association for History and Philosophy of Science has done me a very great honor in conceiving and sponsoring this collection of essays. I owe an enormous debt of gratitude to Robert McLaughlin – my friend and former student – for the diligence, skill, and imagination which he has exercised in editing this volume. As he has noted in his Editorial Preface, there has been a good deal of constructive intellectual interchange among the various contributors (including McLaughlin and myself) during the production of this work, and it has been ably and diplomatically facilitated by the editor. To the authors of the essays, who have contributed their time and talents to this enterprise, I must express my very deepest thanks, for I have benefitted greatly from their work. My hope is that others will find in the fruits of our cooperative venture something of genuine value.

I am grateful to the University of Arizona for granting sabbatical leaves to my wife and me during the year of our visit to Australia, and to the National Science Foundation (U.S.A.) for support of research in all phases of the work connected with my contributions to this volume.

University of Pittsburgh

AUTOBIOGRAPHICAL NOTE

Born in Detroit, Michigan, in 1925, I attended elementary and secondary school there, and took two years' work at Wayne (now Wayne State) University. By the time I entered college, I had decided to become a Methodist minister. In 1944, I went to the University of Chicago Divinity School, but after losing my belief in virtually every tenet of Christianity, as well as all confidence in the social value of religion, I switched my major to philosophy. During that period, Whitehead's metaphysics appealed strongly to me; I took an MA degree in 1947 with a thesis on Whitehead. To my dying day I shall deeply regret the fact that I had not one course from Carnap, who was a member of the Philosophy Department at Chicago at that time.

Following a strong inclination to head west, I went to the University of California at Los Angeles to work toward a doctorate in philosophy. There I encountered Hans Reichenbach, who had a most profound influence upon all of my subsequent work. In 1950, I finished a dissertation on probability and induction, and was granted a PhD.

After receiving my degree, I held a temporary instructorship at UCLA for one year, and in 1951, I took up a post at Washington State College. When I began my career as a professional philosopher, I knew I was seriously interested in philosophy of science, but I had no background in either mathematics or physics. I applied for a fellowship from the Fund for the Advancement of Education to study mathematics, and when it was granted I was able to spend 1953–54 pursuing that work at UCLA. As it happened, Reichenbach died suddenly in the spring of 1953, so in addition to doing mathematics, I spent a good part of the year assisting Maria Reichenbach in preparing various materials, especially *The Direction of Time*, for posthumous publication.

During the following years, I held a visiting appointment at Northwestern University, and in 1953 I went to Brown University to begin my first long-term position. I remained there until 1963. In the spring term of 1959, at the invitation of Stephan Körner, who had visited Brown, I spent four months as a visiting lecturer at the University of Bristol. In 1963, Herbert Feigl arranged a visiting research professorship for me at the Minnesota Center for the Philosophy of Science during the winter and spring quarters. The high point of that visit was the week Feigl, Grover Maxwell, and I spent in

281

Robert McLaughlin (ed.), What? Where? When? Why?, 281–283.
Copyright © 1982 by D. Reidel Publishing Company.

Los Angeles (under the auspices of the Minnesota Center) for talks with
Carnap. For several hours each day we sat under the orange trees in Carnap's
garden discussing his views on probability and induction.

In the autumn of 1963, I began a ten-year stay at Indiana University in the
Department of History and Philosophy of Science. This marked a major turning point for me. Before going to Indiana, I had always been affiliated with
philosophy departments, and most of my published work was on probability
and induction. I considered philosophy of science, as I practiced it, a branch
of philosophy which had important things to say to such other branches of
philosophy as epistemology, metaphysics, and philosophy of religion. Along
with deductive logic and the theory of perception, inductive logic seemed to
me an indispensable ingredient of epistemology. At Indiana, for the first time,
I found myself dealing with graduate students and colleagues who had strong
scientific backgrounds and a fair degree of scientific sophistication. It became
essential for me to treat philosophy of science, not as a discipline looking
inward upon other branches of philosophy, but as one which looks outward
toward the various scientific disciplines. My attempt to make philosophy
relevant to the understanding of science may not have had earth-shaking
significance for any of the sciences, but I regard it as one of the most important philosophical steps I have taken. I came to the conviction, which I still
hold, that philosophy which remains out of contact with other disciplines
runs the great risk of becoming quite sterile.

As a direct result of the reorientation occasioned by my move into an HPS
department, I began working more seriously in the philosophy of space and
time. Coincidentally, in 1962 Hempel published his first systematic essay
on statistical explanation, and this led me, shortly thereafter, to start thinking
seriously about that subject. The intellectual atmosphere at Indiana was
one which greatly facilitated growth, and I shall always be grateful to the
colleagues and graduate students from whom I learned enormously. Two of
them, Clark Glymour and Robert McLaughlin, have contributed essays to
this collection. An added bonus at Indiana was the close connection with
Australian history and philosophy of science, which resulted in lasting friend-
ships with John Clendinnen, Rod Home, and Robert McLaughlin – each of
whom played some significant role in bringing this volume into existence –
and many other people as well. This delightful "Australian connection"
eventually led to a rewarding and enjoyable visit to Melbourne in 1978.

In 1968–69, I spent a year as a visiting professor at the University of
Pittsburgh, where I availed myself of the opportunity to attend Adolf Grünbaum's two-semester graduate course on space and time. My contacts with

Grünbaum over a span of many years have been second only to my contact with Reichenbach in intellectual importance to me.

In 1971, I married Merrilee Ashby, who was at that time teaching philosophy at DePauw University, and I moved to Greencastle, Indiana, where DePauw is located, almost 50 miles from Indiana University in Bloomington. When an opportunity arose for both of us to join the Department of Philosophy at the University of Arizona in 1973, we moved to Tucson. At Arizona, both of us have been able to develop beneficial intellectual relationships with able people in the sciences − anthropology, astronomy, mathematics, and physics, and in the history of science as well − which have done much to enrich our philosophical pursuits. In contrast, the Philosophy Department, which showed great promise in 1973, became intensely uncongenial. We remained at Arizona for eight years, except for our three-month visit to Australia in 1978. Early in 1981 we made the decision to accept positions at the University of Pittsburgh, where I shall be Chairman of the Philosophy Department for the first two or three years of my tenure. The attraction of a group which includes such philosophers of science as Hempel, Grünbaum, Rescher, Sellars, Glymour, Schaffner, and Laudan, as well as first-rate philosophers in other fields, proved irresistible. I look forward to embarking upon a new phase of my career beginning in September of 1981.

During my stay in Melbourne, I began writing a comprehensive monograph on scientific explanation and causality. This manuscript, which now tentatively bears the title, "Scientific Explanation and the Causal Structure of the World," is nearly complete and should soon go to press. Every chapter will bear important marks of the fruitful interactions with philosophers in several Australian universities − some of whom have kindly contributed essays to the present volume. To all of them I wish to offer heartfelt thanks.

NOTES ON CONTRIBUTORS

JOHN CLENDINNEN has taught in the Melbourne University Department of History and Philosophy of Science since the early 1950's. He was born in Melbourne, and after service in the Royal Australian Air Force took a degree in philosophy with one year of science. He completed a part-time PhD in 1972, with a thesis entitled 'Logical Principles of Scientific Inference'. He has published articles in *Mind, Philosophy of Science* and *Synethese*. He was a Visiting Lecturer in 1966 in the Philosophy Department at the University of Pittsburgh, and in 1967 in the History and Philosophy of Science Department at Indiana University. He has worked mainly on scientific inference, but has general interests in the epistemology of science, mathematics and the social sciences. He is married to Inga Clendinnen, who lectures in History at La Trobe University, Melbourne; they have two sons.

JOHN FORGE was born in Northern England. He has attended the Universities of Oxford, Cornell, McGill and London; and has taken a degree in chemistry, two in philosophy and a teaching diploma. He has taught in the Departments of History and Philosophy of Science at the Universities of New South Wales and Wollongong (N.S.W.). One of his main research interests concerns the nature of explanation in physical science and the role of models in physical explanation. He also believes that philosophers of science can make substantial contributions to our understanding of the social relations of science.

CLARK GLYMOUR received his PhD in History and Philosophy of Science from Indiana University in 1969 for a thesis supervised by Wesley Salmon. He was Assistant and then Associate Professor of Philosophy at Princeton University from 1969–1976, and since then has taught at the University of Oklahoma and the University of Illinois at Chicago Circle. He is presently Professor of History and Philosophy of Science at the University of Pittsburgh. Professor Glymour is the co-editor (with J. Earman and J. Stachel) of *Foundations of Space-Time Theories* (University of Minnesota Press, 1977) and is the author of *Theory and Evidence* (Princeton University Press, 1980).

Robert McLaughlin (ed.), What? Where? When? Why?, 285–288.
Copyright © 1982 by D. Reidel Publishing Company.

Australian-born, ROBERT McLAUGHLIN attended Indiana University from 1964 to 1967, where he took his PhD in philosophy of science under the supervision of Wesley Salmon. In Australia he had earlier completed a Master's degree in psychology, and had practised as a psychologist for a number of years, including a triennium with the Royal Australian Navy. He spent the first half of 1967 at the University of Minnesota as Visiting Instructor in Philosophy and Research Associate to Herbert Feigl in the Minnesota Center for Philosophy of Science. He then returned to Sydney as Lecturer, and later Senior Lecturer, in Philosophy at Macquarie University. His main enterprises within philosophy of science embrace the defence of a realist view of scientific theories, and the attempt to explicate a logic of scientific invention. His publications include: 'Educational Psychology: Some Questions of Status' (with R. Precians) in R. J. Selleck (ed.) *Melbourne Studies in Education, 1968–1969* (Melbourne University Press, 1969). Married, with two daughters, he resides permanently in Sydney.

HUGH MELLOR, born in 1938, took his first degree at Cambridge in chemical engineering. He later completed a MSc at the University of Minnesota in chemical engineering with a minor in philosophy; after which he proceeded to a PhD in philosophy at Cambridge. He is presently Lecturer in Philosophy at Cambridge University, and Fellow of Darwin College, Cambridge. In 1975 he was a Visiting Research Fellow at the Australian National University, Canberra; in 1978–1979 he was British Academy Overseas Visiting Research Fellow at the University of California, Berkeley and at Stanford University; and during 1978–1980 he was Radcliffe Trust Fellow. Dr. Mellor was Editor of *The British Journal for Philosophy of Science*, 1968–1970, and is currently Series Editor of *Cambridge Studies in Philosophy*. He has also edited the following works: a new edition of F. P. Ramsey's *Foundations of Mathematics* (Routledge and Kegan Paul, 1978); *Science, Belief and Behaviour* (CUP, 1980); and *Prospects for Pragmatism* (CUP, 1980). He has written sundry articles, and is the author of two books – *The Matter of Chance* (CUP, 1971) and *Real Time* (CUP, 1981).

GRAHAM NERLICH was born in Adelaide, where he took his first degree under J. J. C. Smart, who was then Professor of Philosophy at the University of Adelaide. He pursued graduate studies at Oxford, 1956–1958, where he gained the BPhil. He has taught at the Universities of Leicester (1958–1961); Sydney (1962–1973) – where he attained his first chair; and Adelaide – where he presently holds the Chair of Philosophy. He has visited and lectured at the following universities: Johns Hopkins, Hong Kong (where he was Leverhulme Fellow in 1973), Auckland, Cambridge and Oxford. Professor Nerlich has published papers on a variety of topics within epistemology and the philosophy of logic, but for the last few years he has worked mainly on the philosophy of space and time. He is the author of *The Shape of Space* (CUP, 1976); and is a Fellow of the Australian Academy of the Humanities.

JOHN NORTON graduated as a Bachelor of Engineering (Chemical Engineering) from the University of New South Wales (Sydney) in 1975, and then worked for two years at the Shell refinery, Clyde, N.S.W. After a year travelling, he returned to UNSW in 1978 as a PhD candidate in the School of History and Philosophy of Science, under the supervision of John Saunders. His research deals with the foundations and the history of the general theory of relativity, with a special interest in the emergence, both historically and conceptually, of the general theory from gravitational field theory and the special theory of relativity. He has read papers in this area at the 1980 ANZAAS Conference in Adelaide and the 1980 AAHPSSS Conference in Dunedin, New Zealand. He presently tutors in the history of cosmology courses offered by the School.

JOHN SAUNDERS is now a Senior Consultant in Energy Policy and Resource Planning with the Sydney-based group W. D. Scott & Company, and a Research Associate of the Technology and Society Program at the University of New South Wales (Sydney). After reading mathematical physics at the University of Sussex, he returned to Australia in 1968 as a doctoral scholar in theoretical physics at the Australian National University, and subsequently completed his PhD on the foundations of the special theory of relativity at UNSW in 1972. As a Lecturer, and then Senior Lecturer, in the School of History and Philosophy of Science at UNSW, he was responsible for developing research and teaching programs in the philosophy of science and the conceptual framework of relativity physics and cosmology. As the first Co-ordinator of the Master of Science and Society Course his interests included the past and future contribution of the physics community to problems of modern strategy and defence. He has written and broadcast on a wide range of themes, including the foundations and future of physics, technological innovation and social change, and the methodological problems and strategic implications of energy policy.

D. C. STOVE is Associate Professor in the Department of Traditional and

Modern Philosophy, University of Sydney. He was born in 1927, and is an Australian citizen, married, with two children. His main philosophical interests are in Hume, induction, probability and positivism. He has written many articles, mainly on those subjects, in Australian, British, Canadian and U.S. journals; and two books, *Probability and Hume's Inductive Scepticism* (Oxford UP, 1973), and *Popper and After* (Pergamon Press, forthcoming 1981).

BAS C. VAN FRAASSEN presently teaches at the University of Toronto and the University of Southern California in alternating semesters. He earlier served on the faculties of Yale and Indiana. In 1970–1971 he was a Guggenheim Fellow, and in 1979 a Connaught Fellow. He edited the *Journal of Symbolic Logic*, 1970–1977, and is currently a joint editor of *Pacific Philosophical Quarterly*. Apart from many articles, Professor van Fraassen's writings include the following books: *An Introduction to the Philosophy of Time and Space* (Random House, 1970); *Formal Semantics and Logic* (Macmillan, 1971); *Derivation and Counterexample* (with K. Lambert; Wadsworth-Dickenson, 1972); and *The Scientific Image* (Oxford UP, 1980). Together with E. Beltrametti, he is currently editing a collection of studies on quantum logic.

SALMON BIBLIOGRAPHY

Published Writings of Wesley C. Salmon (to August, 1980)

[1950] 'John Venn's Theory of Induction', PhD dissertation, University of California at Los Angeles.

[1951] 'A Modern Analysis of the Design Argument', *Research Studies of the State College of Washington* 19 (4), 207–220.

[1953a] 'The Frequency Interpretation and Antecedent Probabilities', *Philosophical Studies* 4 (3), 44–48.

[1953b] 'The Uniformity of Nature', *Philosophy and Phenomenological Research* 14 (1), 39–48.

[1955] 'The Short Run', *Philosophy of Science* 22 (3), 214–221.

[1956a] 'Reply to Pettijohn', *Philosophy of Science* 23 (2), 150–152.

[1956b] 'Regular Rules of Induction', *The Philosophical Review* 65 (3), 385–388.

[1957a] 'Should We Attempt to Justify Induction?', *Philosophical Studies* 8 (3), 33–48.

[1957b] 'The Predictive Inference', *Philosophy of Science* 24 (2), 180–190.

[1957c] (with G. Nakhnikian) '"Exists" as a Predicate', *The Philosophical Review* 66 (4), 535–542.

[1957d] Review of Sir Harold Jeffreys, *Scientific Inference*, 2nd edn., *Philosophy of Science* 24 (4), 364–366.

[1957e] Review of Sheldon J. Lachman, *The Foundations of Science*, *Philosophy of Science* 24 (4), 358–359.

[1958] Review of James T. Culbertson, *Mathematics and Logic for Digital Devices*, *Mathematical Reviews* 19 (11), 1200.

[1959a] 'Psychoanalytic Theory and Evidence' in S. Hook (ed.) *Psychoanalysis, Scientific Method, and Philosophy*, New York: New York University Press, pp. 252–267.

[1959b] Review of S. F. Barker, *Induction and Hypothesis*, *The Philosophical Review* 68 (2), 387–388.

[1959c] 'Barker's Theory of the Absolute', *Philosophical Studies* 10 (4), 50–53.

[1959d] Review of Häkan Tornebohm, 'On Two Logical Systems Proposed in the Philosophy of Quantum Mechanics', *Mathematical Reviews* 20 (4), 2278.

[1959e] Review of G. H. von Wright, *The Logical Problem of Induction*, 2nd edn., *Philosophy of Science* 26 (2), 166.

[1959f] Review of Richard von Mises, *Probability, Statistics, and Truth*, 2nd English edn., *Philosophy of Science* 26 (4), 387–388.

[1960a] Review of Hans Reichenbach, *The Philosophy of Space and Time*, *Mathematical Reviews* 21 (5), 3247.

[1960b] Review of Hans Reichenbach, *Modern Philosophy of Science*, *The Philosophical Review* 69 (3), 409–411.

[1960c] Review of J. L. Destouches, 'Physico-logical problems', *Mathematical Reviews* 21 (10), 6319.

[1961a] 'Vindication of Induction', in H. Feigl and G. Maxwell (eds.), *Current Issues in the Philosophy of Science*, New York: Holt, Rinehart and Winston, pp. 245–256.

[1961b] 'Rejoinder to Barker', ibid., pp. 260–262.

[1961c] 'Comments on Barker's "The Role of Simplicity in Explanation"', ibid., pp. 274–276.

[1962] Review of R. Harré, *An Introduction to the Logic of the Sciences*; D. Greenwood, *The Nature of Science*; and V. E. Smith, *The General Science of Nature*, *Isis* 53 (2), 234–235.

[1963a] *Logic*, Englewood Cliffs, N.J.: Prentice-Hall, Inc. Translated into Italian, Japanese, Portuguese, and Spanish.

[1963b] 'On Vindicating Induction', in H. Kyburg and E. Nagel (eds.), *Induction: Some Current Issues* Middletown, Conn.: Wesleyan University Press, pp. 27–41. Also in *Philosophy of Science* 30 (3), 252–261.

[1963c] 'Reply to Black', in Kyburg and Nagel, op. cit., pp. 49–54.

[1963d] 'Inductive Inference', in B. H. Baumrin (ed.), *Philosophy of Science: The Delaware Seminar II*, New York: John Wiley & Sons, pp. 341–370.

[1963e] Review of J. P. Day, *Inductive Probability*, *The Philosophical Review* 72 (3), 392–396.

[1963f] Review of I. J. Good, 'A Causal Calculus I, II', *Mathematical Reviews* 26 (1), 789a–b.

[1964a] Review of R. G. Colodny (ed.), *Frontiers of Science and Philosophy*, *Isis* 55 (2), 215–216.

[1964b] Letter to the Editor, *New York Review of Books*, May 14. Reply to S. Toulmin's review of A. Grünbaum, *Philosophical Problems of Space and Time*.

[1965a] 'The Concept of Inductive Evidence', *American Philosophical Quarterly* 2 (4), 1–6.

[1965b] 'Rejoinder to Barker and Kyburg', ibid., pp. 13–16.

[1965c] 'What Happens in the Long Run?', *The Philosophical Review* 74 (3), 372–378.

[1965d] 'The Status of Prior Probabilities in Statistical Explanation', *Philosophy of Science* 32 (2), 137–146.

[1965e] 'Reply to Kyburg', ibid., pp. 152–154.

[1965f] 'Consistency, Transitivity, and Inductive Support', *Ratio* 7 (2), 167–169.

[1965g] Review of A. Grünbaum, *Philosophical Problems of Space and Time*, *Science* 147 (3659), 724–725.

[1966a] 'The Foundations of Scientific Inference', in R. G. Colodny (ed.), *Mind and Cosmos*, Pittsburgh: University of Pittsburgh Press, pp. 135–275.

[1966b] 'Verifiability and Logic', in P. K. Feyerabend and G. Maxwell (eds.), *Mind, Matter, and Method*, Minneapolis: University of Minnesota Press, pp. 354–376.

[1966c] 'Use, Mention, and Linguistic Invariance', *Philosophical Studies* 17 (1–2), 13–18.

[1967a] *The Foundations of Scientific Inference*, Pittsburgh: University of Pittsburgh Press. Reprint of [1966a], with Addendum 1967 added.

[1967b] 'Carnap's Inductive Logic', *Journal of Philosophy* 64 (21), 725–739.

[1967c] 'Empirical Statements about the Absolute', *Mind* 74 (303), 430–431.

[1967d] Review of R. Carnap, *Philosophical Foundations of Physics*, *Science* 155 (3767), 1235.

[1967e] Review of H. Kyburg, *Probability and the Logic of Rational Belief*, *Philosophy of Science* 34 (3), 283–285.

[1968a] 'Inquiries into the Foundations of Science', in D. L. Arm (ed.), *Vistas in Science*, Albuquerque: University of New Mexico Press, pp. 1–24.

[1968b] 'The Justification of Inductive Rules of Inference', in I. Lakatos (ed.), *The Problem of Inductive Logic*, Amsterdam: North-Holland Publishing Co., pp. 24–43.

[1968c] 'Reply', ibid., pp. 74–97.

[1968d] 'Who Needs Inductive Acceptance Rules', ibid., pp. 139–144.

[1969a] 'Introduction: The Context of these Essays', *Philosophy of Science* 36 (1), 1–4. (With A. Grünbaum; an introduction to 'A Panel Discussion of Simultaneity by Slow Clock Transport in the Special and General Theories of Relativity'.)

[1969b] 'The Conventionality of Simultaneity', *Philosophy of Science* 36 (1), 44–63.

[1969c] 'Partial Entailment as a Basis for Inductive Logic', in N. Rescher (ed.) *Essays in Honor of Carl G. Hempel*, Dordrecht: D. Reidel Publishing Co., pp. 47–82.

[1969d] 'Comment [on R. M. Martin]', in J. Margolis (ed.), *Fact and Existence*, Oxford: Basil Blackwell, pp. 95–97.

[1969e] 'Induction and Intuition: Comments on Ackermann's "Problems" ', in J. W. Davis *et al.* (eds.), *Philosophical Logic*, Dordrecht: D. Reidel Publishing Co., pp. 158–163.

[1969f] 'Foreword to the English Edition' of Hans Reichenbach, *Axiomatization of the Theory of Relativity*, Berkeley and Los Angeles: University of California Press, pp. v–x.

[1970a] *Zeno's Paradoxes* (ed.), Indianapolis: Bobbs-Merrill.

[1970b] 'Statistical Explanation' in R. G. Colodny (ed.), *The Nature and Function of Scientific Theories*, Pittsburgh: University of Pittsburgh Press, pp. 173–232. Reprinted in [1971a].

[1970c] 'Bayes's Theorem and the History of Science' in R. H. Stuewer (ed.) *Minnesota Studies in the Philosophy of Science*, vol. V, Minneapolis: University of Minnesota Press, pp. 68–86.

[1970d] Review of Adolf Grünbaum, *Modern Science and Zeno's Paradoxes*, *Ratio* 12 (2), 178–182.

[1971a] *Statistical Explanation and Statistical Relevance*, Pittsburgh: University of Pittsburgh Press. With contributions by J. G. Greeno and R. C. Jeffrey.

[1971b] 'Determinism and Indeterminism in Modern Science', in J. Feinberg (ed.), *Reason and Responsibility*, 2nd edn., Encino, Calif.: Dickenson Publishing Co., pp. 316–332.

[1971c] 'Explanation and Relevance: Comments on James G. Greeno's "Theoretical Entities in Statistical Explanation" ', in R. C. Buck and R. S. Cohen (eds.) *PSA 1970*, Dordrecht: D. Reidel Publishing Co., pp. 27–39.

[1971d] 'A Zenoesque Problem' (published without title) in Martin Gardner, 'Mathematical Games', *Scientific American* 225 (6), 97–99, Reprinted in [1975a], pp. 49–51.

[1972a] 'Logic', *Encyclopedia Americana* 17, 673–687.

[1972b] 'Law, in Science', *Encyclopedia Americana* 17, 81.

[1972c] 'Postscript to "Probabilities and the Problem of Individuation" by Bas van Fraassen', in S. Luckenbach (ed.) *Probabilities, Problems, and Paradoxes*, Encino, Calif.: Dickenson Publishing Co., pp. 135–138. (Paraphrase, by van Fraassen, of comments on his paper.)

[1973a] *Logic*, 2nd edn., Englewood Cliffs, N.J.: Prentice-Hall, Inc.

[1973b] 'Confirmation', *Scientific American* 228 (5), 75–83.

[1973c] 'Reply to Bradley Efron', *Scientific American* 229 (3), 8–10.

[1973d] 'Reply to Lehman', *Philosophy of Science* 40 (3), 397–402.

[1973e] 'Numerals vs. Numbers', *DePauw Mathnews* 2 (1), 8–11.

[1974a] 'Memory and Perception in *Human Knowledge*', in G. Nakhnikian (ed.), *Bertrand Russell's Philosophy*, London: Gerald Duckworth, pp. 139–167.

[1974b] 'Russell on Scientific Inference *or* Will the Real Deductivist Please Stand Up?', ibid., pp. 183–208.

[1974c] 'Dedication to Leonard J. Savage', in K. F. Schaffner and R. S. Cohen (eds.), *PSA 1972*, Dordrecht: D. Reidel Publishing Co., p. 101.

[1974d] 'Comments on "Hempel's Ambiguity" by J. Alberto Coffa', *Synthese* 28 (4), 165–170.

[1975a] *Space, Time and Motion: A Philosophical Introduction*, Encino, Calif.: Dickenson Publishing Co.

[1975b] 'An Encounter with David Hume', in J. Feinberg (ed.), *Reason and Responsibility*, 3rd edn., Encino, Calif.: Dickenson Publishing Co., pp. 190–208.

[1975c] 'Confirmation and Relevance', in G. Maxwell and R. M. Anderson Jr. (eds.), *Minnesota Studies in the Philosophy of Science*, vol. VI, Minneapolis: University of Minnesota Press, pp. 3–36.

[1975d] 'Theoretical Explanation', in S. Körner (ed.), *Explanation*, Oxford: Basil Blackwell, pp. 118–145.

[1975e] 'Reply to Comments', ibid., pp. 160–184.

[1975f] 'Note on Russell's Anticipations', *Russell*, No. 17, p. 29.

[1976a] 'Clocks and Simultaneity in Special Relativity *or* Which Twin Has the Timex?', in P. Machamer & R. G. Turnbull (eds.), *Motion and Time, Space and Matter*, Columbus: Ohio State University Press, pp. 508–545. Also published as Chapter IV of [1975a].

[1976b] 'Foreword' to Hans Reichenbach, *Laws, Modalities, and Counterfactuals*, Berkeley & Los Angeles: University of California Press, pp. vii–xlii. Also published under the title, 'Laws, Modalities, and Counterfactuals', *Synthese* 35 (2), 191–229, and in [1979a], pp. 655–696.

[1977a] 'Han Reichenbachs Leben und die Tragweite seiner Philosophie', in A. Kamlah and M. Reichenbach (eds.), *Hans Reichenbach: Gesammelte Werke*, vol. I, Wiesbaden: Vieweg, pp. 5–81. (Translated from English by M. Reichenbach.)

[1977b] 'The Philosophy of Hans Reichenbach', *Synthese* 34 (1), 5–88. (Revised version of English original for [1977a].) Also published in [1979a], pp. 1–84.

[1977c] 'Hans Reichenbach', in J. A. Garraty (ed.), *Dictionary of American Biography*, Supplement Five, 1951–1955 New York: Charles Scribners Sons, pp. 562–563.

[1977d] 'Hempel's Conception of Inductive Inference in Inductive-Statistical Explanation', *Philosophy of Science* 44 (2), 180–185.

[1977e] 'Indeterminism and Epistemic Relativization', ibid., pp. 199–202.

[1977f] 'An "At-At" Theory of Causal Influence', ibid., pp. 215–225.

[1977g] 'A Third Dogma of Empiricism' in R. E. Butts and J. Hintikka (eds.), *Basic Problems in Methodology and Linguistics*, Dordrecht: D. Reidel Publishing Co., pp. 149–166.

[1977h] 'Objectively Homogeneous Reference Classes', *Synthese* 36 (4), 399–414.

[1977i] 'An Ontological Argument for the Existence of the Null Set', in Martin Gardner, *Mathematical Magic Show*, New York: Alfred A. Knopf, pp. 32–33.

[1977j] 'The Curvature of Physical Space', in J. S. Earman, C. N. Glymour, and J. Stachel (eds.), *Minnesota Studies in the Philosophy of Science*, vol. VIII, Minneapolis: University of Minnesota Press, pp. 281–302.

[1977k] 'The Philosophical Significance of the One-Way Speed of Light', *Noûs* 11 (3), 253–292.

[1978a] 'Unfinished Business: The Problem of Induction', *Philosophical Studies* 33 (1), 1–19.

[1978b] 'Religion and Science: A New Look at Hume's *Dialogues*', *Philosophical Studies* 33 (2), 143–176.

[1978c] 'Why Ask, "Why?"'? – An Inquiry Concerning Scientific Explanation', *Proceedings and Addresses of the American Philosophical Association* 51 (6), 683–705. Also published in [1979a], pp. 403–425.

[1978d] 'Hans Reichenbach: A Memoir', in M. Reichenbach and R. S. Cohen (eds.), *Hans Reichenbach: Selected Writings, 1909–1953*, Dordrecht: D. Reidel Publishing Co., pp. 69–77.

[1979a] (ed.) *Hans Reichenbach: Logical Empiricist*, Dordrecht: D. Reidel Publishing Co. Large portions were previously published in *Synthese* 34 (1–3) and 35 (1–2).

[1979b] 'Postscript: Laws in Deductive-Nomological Explanation – An Application of the Theory of Nomological Statements', ibid., pp. 691–694.

[1979c] (With M. H. Salmon), 'Alternative Models of Scientific Explanation', *American Anthropologist* 81 (1), 61–74.

[1979d] 'Experimental Atheism', *Philosophical Studies* 35 (1), 101–104.

[1979e] 'Philosopher in a Physics Course', *Teaching Philosophy* 2 (2), 139–146.

[1979f] 'Propensities: A Discussion-Review', *Erkenntnis* 14 (2), 183–216.

[1979g] 'Informal Analytic Approaches to the Philosophy of Science', in P. D. Asquith & H. E. Kyburg Jr. (eds.), *Current Research in Philosophy of Science*, East Lansing, Mich.: Philosophy of Science Assn., pp. 3–15.

[1980a] 'Probabilistic Causality', *Pacific Philosophical Quarterly* 61 (1–2), 50–74.

[1980b] *Space, Time, and Motion: A Philosophical Introduction*, 2nd edn., Minneapolis: University of Minnesota Press.

[1980c] 'John Venn's *Logic of Chance*' in J. Hintikka & D. Gruender (eds.) *Probabilistic Thinking, Thermodynamics, and the Interaction of History and Philosophy of Science*, vol. II, Dordrecht: D. Reidel Publishing Co., pp. 125–138.

[1980d] 'Robert Leslie Ellis and the Frequency Theory', ibid., pp. 139–143.

[1980e] 'Wissenschaft, Grundlagen der,' in J. Speck, ed., *Handbuch Wissenschafts-theoretischer Begriffe*. Göttingen: Vandenhoeck & Ruprecht, pp. 752–757. [Translated from English by the editor.]

Forthcoming:

'Causality: Production and Propagation' in Peter D. Asquith and Ronald N. Giere, eds., *PSA 1980*, vol. II, East Lansing, Mich.: Philosophy of Science Assn.

'Rational Prediction', *British Journal for the Philosophy of Science*.

'Causality in Archaeological Explanation', Proceedings of a 1980 Meeting of the Theoretical Archaeology Group, University of Southampton, organized by A. C. Renfrew.

'In Praise of Relevance', Proceedings of a Conference on the Teaching of Logic, Carnegie-Mellon University, organized by Preston Covey.

REFERENCES

Achinstein, P. [1970] 'Inference to Scientific Laws', in R. H. Stuewer (ed.), *Minnesota Studies in Philosophy of Science*, Vol. V, Minneapolis: Univ. of Minnesota Press, pp. 87–111.

Atkins, P. W. [1970] *Molecular Quantum Mechanics*, Vol. I, Oxford: Oxford University Press.

Bergmann, P. G. [1976] *Introduction to the Theory of Relativity*, New York: Dover.

Black, M. [1954] *Problems of Analysis*, Ithaca: Cornell UP.

Blackburn, S. [1980] 'Opinions and Chances', in D. H. Mellor (ed.), *Prospects for Pragmatism*, Cambridge: Cambridge UP.

Bowman, P. 'Einstein's Second Treatment of Simultaneity', *Philosophy of Science Association* 1, 71–81.

Braithwaite, R. B. [1953] *Scientific Explanation*, London: Cambridge UP.

Bridgman, P. W. [1962] *A Sophisticate's Primer of Relativity*, Middletown, Conn.: Wesleyan UP.

Brody, B. [1975] 'The Reduction of Teleological Sciences', *American Philosophical Quarterly* 12 (1), 69–76.

Bub, J. [1974] *The Interpretation of Quantum Mechanics*, Dordrecht: Reidel.

Bunge, M. [1973] *Philosophy of Physics*, Dordrecht: Reidel.

Cajori, F. (ed.) [1947] *Sir Isaac Newton's Mathematical Principles of Natural Philosophy and his System of the World*, Berkeley & Los Angeles: Univ. of California Press.

Cantrell, C. D. and Scully, M. O. [1978] 'The EPR Paradox Revisited', *Physics Reports* 43 (13), 499–508.

Carnap, R. [1936–7] 'Testability and Meaning', *Philosophy of Science* 3, 420–468; 4, 1–40.

Carnap, R. [1950], [1962] *Logical Foundations of Probability* (1st, 2nd edns.), Chicago: Univ. of Chicago Press.

Carnap, R. [1952] *The Continuum of Inductive Methods*, Chicago: Univ. of Chicago Press.

Carnap, R. [1963] 'Replies and Systematic Expositions', in Schilpp (ed.) [1963], pp. 859–1013.

Cattell, R. [1978] *The Scientific Use of Factor Analysis*, New York: Plenum Press.

Clendinnen, F. J. [1966] 'Induction and Objectivity', *Philosophy of Science* 33 (3), 215–229.

Clendinnen, F. J. [1977] 'Inference, Practice and Theory', *Synthese* 34 (1), 89–132.

Clendinnen, F. J. [1981] 'Rational Expectation and Simplicity', this volume pp. 1–25.

De Finetti, B. [1937], [1964] 'Foresight: Its Logical Laws, Its Subjective Source', translation in H. E. Kyburg and H. E. Smokler (eds.), *Studies in Subjective Probability*, New York: Wiley, pp. 93–158.

d'Espagnat, B. [1979] 'Quantum Theory and Reality', *Scientific American* 241 (5), 158–181.

Dirac, P. A. M. [1958] *The Principles of Quantum Mechanics* (4th Edn.), Oxford: Oxford UP.

Dirac, P. A. M. [1978] *Directions in Physics*, New York: John Wiley & Sons.

Earman, J. [1971] 'Laplacian Determinism, Or Is This Any Way to Run a Universe', *Journal of Philosophy* 68 (21), 729–744.

Earman, J., Glymour, C., and Stachel, J. (eds.) [1977] *Minnesota Studies in the Philosophy of Science*, Vol. VIII, Minneapolis: Univ. of Minnesota Press.

Einstein, A. [1905] 'On the Electrodynamics of Moving Bodies', in Einstein *et al.* [1923], [1952], pp. 35–65.

Einstein, A. [1907] 'Uber das Relativitätsprinzip und die aus demselben gezogenen Folgerungen', *Jahrbuch der Radioaktivität* 4, 411–462; 5, 98–99.

Einstein, A. [1911] 'Die Relativitätstheorie', *Naturforschende Gesellschaft, Vierteljahresschrift* (Zurich) 56, 1–14.

Einstein, A. [1916] 'Nachruf auf Mach', *Physikalische Zeitschrift* 17, 101–4.

Einstein, A. [1917], [1977] *Relativity. The Special and the General Theory*, London: Methuen.

Einstein, A. [1921], [1976] *The Meaning of Relativity*, London: Chapman and Hall.

Einstein, A. [1923], [1952] *The Principle of Relativity*, New York: Dover.

Einstein, A. [1936] 'Physics and Reality', *Journal of the Franklin Institute* 221, 349–382.

Einstein, A. [1948] 'Relativity: Essence of the Theory of Relativity', *American People's Encyclopedia*, Vol. 16, Chicago: Spencer Press.

Einstein, A. [1949] 'Autobiographical Notes', in P. A. Schilpp (ed.), *Albert Einstein: Philosopher-Scientist*, New York: Harper, pp. 1–95.

Ellis, B. [1966] *Basic Concepts in Measurement*, Cambridge: Cambridge UP.

Ellis, B. [1971] 'On Conventionality and Simultaneity', *Australasian Journal of Philosophy* 49 (2), 177–203.

Ellis, B. and Bowman, P. [1967] 'Conventionality in Distant Simultaneity', *Philosophy of Science* 34 (2), 116–136.

Feigl, H. [1950] 'De Principiis Non Disputandum', in M. Black (ed.), *Philosophical Analysis*, Englewood Cliffs, N. J.: Prentice-Hall Inc., pp. 113–147.

Feigl, H. [1970] 'The "Orthodox" View of Theories: Remarks in Defense as well as Critique', in M. Radner and S. Winokur (eds.), [1970], pp. 3–16.

Feyerabend, P. K. [1975] *Against Method*, London: New Left Books.

Feynman, R. P. [1949a] 'The Theory of Positrons', *Physical Review* 76, 749. Reprinted in Schwinger [1958], and in Feynman [1961a].

Feynman, R. P. [1949b] 'Space-Time Approach to Quantum Electrodynamics', *Physical Review* 76, 769. Reprinted in Schwinger [1958], and in Feynman [1961a].

Feynman, R. P. [1950] 'Mathematical Formulation of the Quantum Theory of Electromagnetic Interaction', *Physical Review* 80, 440. Reprinted in Schwinger [1958].

Feynman, R. P. [1961a] *Quantum Electrodynamics*, New York: Benjamin.

Feynman, R. P. [1961b] *Theory of Fundamental Processes*, New York: Benjamin.

Frankfurt, H. [1958] 'Peirce's Notion of Abduction', *The Journal of Philosophy* 55 (14), 593–597.

Freud, S. [1924] *Collected Papers* (trans. Joan Riviere), London: The International Psycho-Analytical Press.

Friedman, M. [1974] 'Explanation and Scientific Understanding', *The Journal of Philosophy* 71 (1), 5–19.

Friedman, M. [1977] 'Simultaneity in Newtonian Mechanics and Special Relativity', *Minnesota Studies in Philosophy of Science*, Vol VIII, Minneapolis: Univ. of Minnesota Press, pp. 403–432.

Gardner, M. [1976] *The Incredible Dr. Matrix*, New York: Scribners.

Gardner, M. R. [1979] 'Realism and Instrumentalism in 19th Century Atomism', *Philosophy of Science* 46 (1), 1–34.

Giannoni, C. [1978] 'Relativistic Mechanics and Electrodynamics Without One-Way Velocity Assumptions', *Philosophy of Science* 45 (1), 17–46.

Glymour, C. [1980] *Theory and Evidence*, Princeton: Princeton UP.

Glymour, C. (forthcoming) 'Explanation, Tests, Unity and Necessity', *Noûs*.

Goodman, N. [1965] *Fact, Fiction and Forecast* (2nd edn.), Indianapolis: Bobbs-Merrill.

Greeno, J. G. [1970] 'Evaluation of Statistical Hypotheses Using Information Transmitted', *Philosophy of Science* 37 (2), 279–293.

Greeno, J. G. [1971] 'Explanation and Information', in Salmon *et al.*, [1971], pp. 89–104.

Grünbaum, A. [1955] 'Logical and Philosophical Foundations of the Special Theory of Relativity', *American Journal of Physics* 23, 450–464.

Grünbaum, A. [1968] *Modern Science and Zeno's Paradoxes*, London: George Allen & Unwin Ltd.

Grünbaum, A. [1969] 'Simultaneity by Slow Clock Transport in the Special Theory of Relativity', *Philosophy of Science* 36 (1), 5–43.

Grünbaum, A. [1973] *Philosophical Problems of Space and Time* (2nd enlarged edn.), Dordrecht: Reidel.

Hacking, I. [1965] *Logic of Statistical Inference*, Cambridge: Cambridge UP.

Hacking, I. [1968] 'One Problem About Induction', in Lakatos (ed.) [1968], pp. 44–59.

Hanson, N. R. [1961] 'Is There a Logic of Scientific Discovery?', in H. Feigl and G. Maxwell (eds.), *Current Issues in the Philosophy of Science*, New York: Holt, Rinehart & Winston, pp. 20–35.

Hanson, N. R. [1965] *Patterns of Discovery*, Cambridge: Cambridge UP.

Hanson, N. R. [1967a] 'The Genetic Fallacy Revisited' *American Philosophical Quarterly* 4 (2), 101–113.

Hanson, N. R. [1967b] 'An Anatomy of Discovery', *Journal of Philosophy* 64 (11), 321–352.

Hempel, C. G. [1962a] 'Deductive-Nomological vs. Statistical Explanation', in H. Feigl and G. Maxwell (eds.), *Minnesota Studies in the Philosophy of Science*, Vol. III, Minneapolis: University of Minnesota Press, pp. 98–169.

Hempel, C. G. [1962b] 'Explanation in Science and in History', in R. G. Colodny (ed.), *Frontiers in Science and Philosophy*, Pittsburgh: Univ. of Pittsburgh Press, pp. 7–33.

Hempel, C. G. [1965a] *Aspects of Scientific Explanation* and Other Essays in the Philosophy of Science, New York: The Free Press.

Hempel, C. G. [1965b] 'Aspects of Scientific Explanation', in C. G. Hempel, [1965a], pp. 331–496.

Hempel, C. G. [1966] *Philosophy of Natural Science*, Englewood Cliffs: Prentice-Hall.

Hempel, C. G. [1977] *Aspekte wissenschaftlicher Erklärung* Berlin: Walter de Gruyter.

Hempel, C. G. and Oppenheim, P. [1948] 'Studies in the Logic of Explanation', *Philosophy of Science* 15 (2), 135–175.

Hendel, C. W. (ed.) [1955] *An Enquiry Concerning Human Understanding* by David Hume (includes Hume's "Abstract" as a supplement), Indianapolis: Bobbs-Merrill.

Hesse, Mary B. [1966] *Models and Analogies in Science*, Notre Dame, Ind.: Univ. of Notre Dame Press.

Hesse, Mary B. [1974] *The Structure of Scientific Inference*, London: Macmillan.

Hintikka, J. [1966] 'A Two-Dimensional Continuum of Inductive Methods', in Hintikka and Suppes (eds.), [1966], pp. 113–132.

Hintikka, J. and Suppes, P. (eds.) [1966] *Aspects of Inductive Logic*, Amsterdam: North Holland Pub. Co.

Hoffman, B. [1947] *The Strange Story of the Quantum*, London: Pelican.

Holton, G. [1978] *The Scientific Imagination: Case Studies*, Cambridge: Cambridge UP.

Holton, G. and Brush, S. [1973] *Introduction to Concepts and Theories in Physical Science*, 2nd edn., Reading, Mass.: Addison-Wesley Pub. Co.

Home, R. W. (ed.) [1982] *Science Under Scrutiny*: The Place of History and Philosophy of Science (Australasian Studies in History and Philosophy of Science, Vol. II), Dordrecht: D. Reidel.

Hume, David [1739–40], [1888] *A Treatise of Hume Nature*, See Selby-Bigge (ed.), [1888].

Hume, David [1955] 'An Abstract of *A Treatise of Human Nature*', See Hendel (ed.), [1955].

Hume, David [1777], [1955] *An Enquiry Concerning Human Understanding*, See Hendel (ed.), [1955].

Janis, A. I. [1969] 'Synchronism by Slow Transport of Clocks in Non-inertial Frames of Reference', *Philosophy of Science* 36 (1), 74–81.

Jeffrey, R. C. [1971] 'Statistical Explanation vs. Statistical Inference', in Salmon *et al.*, ([1971], pp. 19–28); first published in Rescher (ed.), ([1969], pp. 104–113).

Jeffrey, R. C. [1980] 'How is it Reasonable to Base Preferences on Estimates of Chance?' in D. H. Mellor (ed.), *Science, Belief and Behaviour*, Cambridge: Cambridge UP.

Jones, R. [1977] 'Causal Anomalies and the Completeness of Quantum Theory', *Synthese* 35 (1), 41–78. Reprinted in Salmon (ed.) [1979a].

Kelvin, Lord (William Thomson) [1884] *Notes of Lectures on Molecular Dynamics and the Wave Theory of Light*, Baltimore: Johns Hopkins UP.

Keynes, J. M. [1921] *A Treatise on Probability*, London: Macmillan.

Koestler, A. [1959] *The Sleepwalkers*, London: Hutchinson.

Koestler, A. [1975] *The Act of Creation*, London: Picador-Pan Books Ltd.

Körner, S. (ed.) [1957] *Observation and Interpretation*, London: Butterworth.

Kripke, S. [1971] 'Identity and Necessity', in M. K. Munitz (ed.), *Identity and Individuation*, New York: New York UP.

Kuhn, T. S. [1962] *The Structure of Scientific Revolutions*, Chicago: Univ. of Chicago Press.

Kyburg, H. E. [1978] 'Subjective Probability: criticisms, reflections and problems' *Journal of Philosophical Logic* 7, 157–180.

Lakatos, I. (ed.) [1968] *The Problem of Inductive Logic*, Amsterdam: North Holland Pub. Co.

Lakatos, I. [1976] *Proofs and Refutations*, Cambridge: Cambridge UP.

Lakatos, I. [1970], [1978a] 'Falsification and the Methodology of Scientific Research Programmes', in Worrall & Currie (eds.), [1978], pp. 8–101.

Lakatos, I. [1970], [1978b] 'History of Science and Its Rational Reconstructions' in
    Worrall and Currie (eds.), [1978], pp. 102–138.
Laplace, P. S., Marquis de [1951] *A Philosophical Essay on Probabilities* (transl. by
    F. W. Truscott and F. L. Emory), New York: Dover Publications.
Laudan, L. [1980] 'Why Was the Logic of Discovery Abandoned?', in Nickles (ed.),
    [1980], pp. 173–183.
Levi, I. [1977] 'Subjunctives, Dispositions and Chances', *Synthese* 34, 423–455.
Lewis, D. [1973] *Counterfactuals*, Cambridge, Mass.: Harvard UP.
Lewis, D. [1979] 'A Subjectivist's Guide to Objective Chance', in R. C. Jeffrey (ed.),
    Studies in Inductive Logic and Probability, Vol. 2, Berkeley: Univ. of California
    Press.
Lewis, D. [forthcoming] 'Causal Explanation'.
Lyon, A. R. [1970] *Dealing with Data*, Oxford: Pergamon.
McLaughlin, R. M. [1967] *Theoretical Entities and Philosophical Dualisms: A Critique
    of Instrumentalism*, Ann Arbor, Michigan: University Microfilms Inc.
Malament, D. [1977] 'Causal Theories of Time and the Conventionality of Simultaneity',
    Noûs 11 (3), 293–300.
Meehl, P. E. [1970] 'Psychological Determinism and Human Rationality: A Psycholo-
    gist's Reaction to Professor Karl Popper's "Of Clouds and Clocks" ', in Radner and
    Winokur (eds.), [1970], pp. 310–372.
Mellor, D. H. [1971] *The Matter of Chance*, Cambridge: Cambridge UP.
Mellor, D. H. [1973] 'Materialism and Phenomenal Qualities II', *Aristotelian Society
    Supplementary Volume* 47, pp. 107–119.
Mellor, D. H. [1974] 'In Defense of Dispositions', *Philosophical Review* 83, 157–181.
Mellor, D. H. [1976] 'Probable Explanation', *Australasian Journal of Philosophy* 54 (3),
    231–241.
Mellor, D. H. [1980] 'Consciousness and Degrees of Belief', in D. H. Mellor (ed.),
    Prospects for Pragmatism, Cambridge: Cambridge UP.
Musgrave, A. [1982] 'Facts and Values in Science Studies', in Home (ed.), [1982].
Nerlich, G. [1976] *The Shape of Space*, Cambridge: Cambridge UP.
Nerlich, G. [1979] 'Is Curvature Intrinsic to Physical Space?' *Philosophy of Science* 46
    (3), 439–458.
Nickles, T. (ed.) [1980] *Scientific Discovery, Logic, and Rationality*. (Boston Studies in
    Philosophy of Science, Vol. 56), Dordrecht: Reidel.
Nye, Mary Jo [1972] *Molecular Reality*, London: Macdonald; New York: American
    Elsevier.
Panofsky, W. K. H. and Phillips, M. [1972] *Classical Electricity and Magnetism*, Reading,
    Mass.: Addison Wesley.
Peirce, C. S. [1931] *Collected Papers*, Cambridge, Mass.: Harvard UP.
Perrin, J. [1916] *Atoms*, London: Constable.
Perry, J. [1979] 'The Problem of the Essential Indexical', *Noûs* 13 (1), 3–21.
Poincaré, H. [1898] 'La Mèsure du Temps', *Revue de Metaphysique et de Morale* 6,
    1.
Poincaré, H. [1902] [1952] *Science and Hypothesis*, New York: Dover.
Popper, K. R. [1934] *Logik der Forschung* (translated and expanded as Popper [1959]),
    Vienna: Springer.
Popper, K. R. [1957] 'The Propensity Interpretation of the Calculus of Probability,
    and the Quantum Theory', in S. Körner (ed.) [1957].

Popper, K. R. [1959] *The Logic of Scientific Discovery*, London: Hutchinson.

Popper, K. R. [1968] 'Theories, Experience and Probabilistic Intuitions', in I. Lakatos (ed.), [1968], pp. 285–303.

Post, H. R. [1971] 'Correspondence, Invariance and Heuristics', *Studies in History and Philosophy of Science* 2 (3), 213–255.

Radcliffe-Brown, A. R. [1952], [1965] *Structure and Function in Primitive Society*, New York: The Free Press.

Radcliffe-Brown, A. R. [1933], [1967] *The Andaman Islanders*, New York: The Free Press.

Radner, M. and Winokur, S. (eds.) [1970] *Minnesota Studies in Philosophy of Science*, Vol. IV, Minneapolis: Univ. of Minnesota Press.

Ramsey, F. P. [1931] *The Foundations of Mathematics*, London: Routledge and Kegan Paul.

Ramsey, F. P. [1926], [1978] 'Truth and Probability', in D. H. Mellor (ed.) [1978] *Foundations*, London: Routledge and Kegan Paul, pp. 58–100.

Ramsey, F. P. [1929], [1978] 'Knowledge' in ibid. pp. 126–127.

Reichenbach, H. [1952] 'Planetenuhr und Einsteinsche Gleichzeitigheit', *Zeitschrift für Physik* 33 (8), 628–634.

Reichenbach, H. [1927], [1958] *Philosophy of Space and Time*, New York: Dover.

Reichenbach, H. [1938], [1961] *Experience and Prediction*, Chicago: Phoenix Books.

Reichenbach, H. [1944] *Philosophic Foundations of Quantum Mechanics*, Berkeley and Los Angeles: Univ. of California Press.

Reichenbach, H. [1949] *The Theory of Probability*, Berkeley and Los Angeles: Univ. of California Press.

Reichenbach. H. [1956] *The Direction of Time*, Berkeley and Los Angeles: Univ. of California Press.

Rescher, N. (ed.) [1969] *Essays in Honor of Carl G. Hempel*, Dordrecht: D. Reidel.

Salmon, W. C. *See separate Salmon Bibliography*.

Salmon, W. C., Jeffrey, R. C., and Greeno, J. G. [1971] *Statistical Explanation and Statistical Relevance*, Pittsburgh: Univ. of Pittsburgh Press.

Saunders, S. D. [1972] 'A Study of Creativity in Scientific Investigation', Unpublished MA (Hons.) Thesis, The University of New South Wales.

Scadron, M. D. [1979] *Advanced Quantum Theory*, New York: Springer-Verlag.

Schilpp, P. A. (ed.) [1949] *Albert Einstein: Philosopher-Scientist* (2 vols.), New York: Harper Torchbooks.

Schilpp, P. A. (ed.) [1963] *The Philosophy of Rudolf Carnap*, La Salle, Ill.: Open Court.

Schlesinger, G. [1963] *Methods in the Physical Sciences*, London: Routledge and Kegan Paul.

Schlesinger, G. [1974] *Confirmation and Confirmability*, Oxford: Clarendon Press.

Schwinger, J. (ed.) [1958] *Selected Papers on Quantum Electrodynamics*, New York: Dover.

Scribner, C. [1963] 'Mistranslation of a Passage in Einstein's Original Paper on Relativity', *American Journal of Physics* 31, 398.

Scriven, M. [1959] 'Explanation and Prediction in Evolutionary Theory', *Science* 130 (3374), 477–482.

Seelig, C. [1956] *Albert Einstein. A Documentary Biography*, London: Staples Press Ltd.

Selby-Bigge, L. A. (ed.) [1888] *A Treatise of Hume Nature* by David Hume, Oxford: Clarendon Press.

Shubnikov, A. V. and Kopstik, V. A. [1974] *Symmetry in Science and Art* (transl. from Russian by E. D. Archard; edited by D. Harker), New York: Plenum Press.

Simon, H. A. [1977] *Models of Discovery*, Dordrecht: Reidel.

Skyrms, B. [1962] 'Falsifiability in the Logic of Experimental Tests', *Estratto Rivista Methodos* **XIV**:

Skyrms, B. [1980] *Causal Necessity*, New Haven: Yale UP.

Smart, J. J. C. [1963] *Philosophy and Scientific Realism*, London: Routledge and Kegan Paul.

Smith, N. Kemp [1941] *The Philosophy of David Hume*, London: Macmillan.

Sober, E. [1975] *Simplicity*, Oxford: Clarendon Press.

Spearman, C. [1927] *The Abilities of Man*, New York: Macmillan.

Stove, D. C. [1973] *Probability and Hume's Inductive Scepticism*, Oxford: Clarendon Press.

Strawson, P. F. [1952] *Introduction to Logical Theory*, London: Methuen.

Stuewer, R. (ed.) [1970] *Minnesota Studies in Philosophy of Science*, Vol. V, Minneapolis: Univ. of Minnesota Press.

Suppes, P. [1974] 'Popper's Analysis of Probability in Quantum Mechanics', in P. A. Schilpp (ed.), *The Philosophy of Karl Popper*, Vol. 2, La Salle, Illinois: Open Court, pp. 760–774.

Suppes, P. and Zanotti, M. [1976] 'On the Determinism of Hidden Variable Theories and Conditional Statistical Independence of Observables', in P. Suppes (ed.), *Logic and Probability in Quantum Mechanics*, Dordrecht: Reidel pp. 445–455.

Taylor, E. F. and Wheeler, J. A. [1966] *Spacetime Physics*, San Francisco: W. H. Freeman & Co.

Van Fraassen, B. [1969] 'Conventionality in the Axiomatic Foundations of the Special Theory of Relativity', *Philosophy of Science* **36** (1), 64–73.

Van Fraassen, B. [1972] 'Probabilities and the Principle of Individuation', in S. Luckenbach (ed.), *Probability: Problems and Paradoxes*, Encino, California: Dickenson-Wadsworth, pp. 121–138.

Van Fraassen, B. [1974] 'The Einstein-Podolsky-Rosen Paradox', *Synthese* **29**, 291–309.

Van Fraassen, B. [1977] 'The Pragmatics of Explanation', *American Philosophical Quarterly* **14** (2), 143–150.

Van Fraassen, B. [1980] *The Scientific Image*, Oxford: Oxford UP.

Venn, J. [1866], [1962] *The Logic of Chance* (4th edn., 1962), New York: Chelsea Pub. Co.

Von Wright, G. H. [1971] *Explanation and Understanding*, London: Routledge and Kegan Paul.

Waismann, F. [1951] 'Verifiability', in A. G. N. Flew (ed.), *Logic and Language – First Series*, Oxford: Basil Blackwell, pp. 117–144.

Weinberg, S. [1979] 'Einstein and Spacetime: Then and Now', *Bulletin of the American Academy of Arts and Sciences* **33** (2).

Whittaker, E. T. [1949] *From Euclid to Eddington*, Cambridge: Cambridge UP.

Whittaker, E. T. [1960] *A History of the Theories of Aether and Electricity*, New York: Harper.

Wichmann, E. H. [1967] *Quantum Physics*, New York: McGraw Hill.

Winnie, J. [1970] 'Special Relativity Without One-Way Velocity Assumptions', *Philosophy of Science* **37** (1), 81–99; **37** (2), 223–238.

Winnie, J. [1977a] 'The Causal Theory of Space-Time', in Earman, Glymour and Stachel
   (eds.) [1977], pp. 134–205.
Winnie, J. [1977b] 'Symposium on Space and Time', *Noûs* 11 (3), 207–209.
Worrall, J. and Currie, G. (eds.) [1978] *The Methodology of Scientific Research Pro-
   grammes* (Lakatos Philosophical Papers, Vol. 1), Cambridge: Cambridge UP.
Wright, L. [1976] *Teleological Explanations*, Berkeley: Univ. of California Press.

INDEX OF NAMES

303

INDEX OF SUBJECTS